Microbiology Experiments:
A Health Science Perspective

Sixth Edition

John Kleyn

Mary Bicknell
University of Washington

Boston   Burr Ridge, IL   Dubuque, IA   New York   San Francisco   St. Louis
Bangkok   Bogotá   Caracas   Kuala Lumpur   Lisbon   London   Madrid   Mexico City
Milan   Montreal   New Delhi   Santiago   Seoul   Singapore   Sydney.   Taipei   Toronto

Higher Education

MICROBIOLOGY EXPERIMENTS: A HEALTH SCIENCE PERSPECTIVE, SIXTH EDITION

Published by McGraw-Hill, a business unit of The McGraw-Hill Companies, Inc., 1221 Avenue of the
Americas, New York, NY 10020. Copyright © 2009 by The McGraw-Hill Companies, Inc. All rights reserved.
Previous editions © 2007, 2004, 2001, 1999, and 1995. No part of this publication may be reproduced or
distributed in any form or by any means, or stored in a database or retrieval system, without the prior written
consent of The McGraw-Hill Companies, Inc., including, but not limited to, in any network or other electronic
storage or transmission, or broadcast for distance learning.

Some ancillaries, including electronic and print components, may not be available to customers outside
the United States.

This book is printed on recycled, acid-free paper containing 10% postconsumer waste.

3 4 5 6 7 8 9 0 QDB/QDB 11

ISBN 978–0–07–299549–7
MHID 0–07–299549–1

Publisher: *Michelle Watnick*
Senior Sponsoring Editor: *James F. Connely*
Director of Development: *Kristine Tibbetts*
Project Coordinator: *Mary Jane Lampe*
Senior Production Supervisor: *Laura Fuller*
Associate Design Coordinator: *Brenda A. Rolwes*
Cover Designer: *Studio Montage, St. Louis, Missouri*
(USE) Cover Image: *Dennis Kunkel Microscopy, Inc.*
Senior Photo Research Coordinator: *John C. Leland*
Photo Research: *David Tietz/Editorial Image, LLC*
Compositor: *S4Carlisle Publishing Services*
Typeface: *11.5/13 Goudy*
Printer: *Quad/Graphics Dubuque, IA*

Some of the laboratory experiments included in this text may be hazardous if materials are handled
improperly or if procedures are conducted incorrectly. Safety precautions are necessary when you are
working with chemicals, glass test tubes, hot water baths, sharp instruments, and the like, or for any
procedures that generally require caution. Your school may have set regulations regarding safety
procedures that your instructor will explain to you. Should you have any problems with materials or
procedures, please ask your instructor for help.

www.mhhe.com

Dedicated to the memory of my husband, who
never tired of biological questions.

Dr. James Noel Bicknell
1938–2008

Contents

Preface

To the Student

A microbiology laboratory is valuable because it actually gives you a chance to see and study microorganisms firsthand. In addition, it provides you with the opportunity to learn the special techniques used to study and identify these organisms. The ability to make observations, record data, and analyze results is useful throughout life.

It is very important to read the scheduled exercises before coming to class so that class time can be used efficiently. It is helpful to ask yourself the purpose of each step as you are reading and carrying out the steps of the experiment. Sometimes it will be necessary to read an exercise several times before it makes sense.

Conducting experiments in microbiology laboratories is particularly gratifying because the results can be seen in a day or two (as opposed, for instance, to plant genetics laboratories). Opening the incubator door to see how your cultures have grown and how the experiment has turned out is a pleasurable moment. We hope you will enjoy your experience with microorganisms as well as acquire skills and understanding that will be valuable in the future.

To the Instructor

The manual includes a wide range of exercises—some more difficult and time-consuming than others. Usually more than one exercise can be done in a two-hour laboratory period. In these classes, students can actually see the applications of the principles they have learned in the lectures and text. We have tried to integrate the manual with the text *Microbiology: A Human Perspective,* Sixth Edition by Eugene Nester et al.

The exercises were chosen to give students an opportunity to learn new techniques and to expose them to a variety of experiences and observations. We did not assume that the school or department has a large budget; thus, exercises were written to

use as little expensive media and equipment as possible. The manual contains more exercises than can be done in one course so that instructors will have an opportunity to select the appropriate exercises for their particular students and class.

An online Instructor's Manual is available from the publisher. It lists equipment, cultures, media, and reagents needed for each exercise and has extensive information for storing cultures and making media. The Notes to the Instructor section gives suggestions for preparing and presenting the laboratory sessions, and there is a section of questions that can be used to supplement those in the student manual. **We highly recommend that the instructors utilize this additional aid.** Contact your local McGraw-Hill representative for the URL and password for this site.

Revisions to the Sixth Edition include a new exercise on acne and an updated identification of the pathogenic *streptococci*. Some of the exercises have been rearranged to give a more logical introduction to techniques, and others have been simplified. The Staphylococcus Carrier Study, an exercise that has potential health risks, has been eliminated.

If a student has a special interest in microbiology and would like to do independent work, four special projects are included. Two are available online: the isolation of *Deinococcus,* an organism exceptionally resistant to radiation; and the isolation of an organism utilizing methanol from plant leaves. You can access these special projects at www.mhhe.com/Nester6e. Isolation techniques for two other interesting organisms, petroleum-utilizing bacteria (important in oil-polluted water) and bioluminescent bacteria, are in the back of the manual. The luminescent bacteria are particularly intriguing. Not only are they an example of quorum sensing, but they are exciting to view in a dark room. Also, these bacteria can be easily isolated by the whole class.

We hope these changes are helpful and that the manual contributes to the students' understanding of microbiology. We also hope both students and instructors enjoy their experience with a very interesting group of organisms.

Acknowledgments

We would like to acknowledge the contributions of the lecturers in the Department of Microbiology at the University of Washington who have thoughtfully honed laboratory exercises over the years until they really work. These include Carol Laxson, Mona Memmer, Janis Fulton, and Mark Chandler. Special thanks to Dale Parkhurst for his expert knowledge of media. We also thank the staff of the University of Washington media room, headed by Jim Calbreath, for their expertise and unstinting support.

We also want to thank Eugene and Martha Nester, Nancy Pearsall, Denise Anderson and Evans Roberts for their text *Microbiology: A Human Perspective*. This text was the source of much of the basic conceptual material and figures for our laboratory manual.

We gratefully acknowledge the contributions of Anna Oller, from the University of Central Missouri.

Her explanation of current techniques and concepts was very valuable in our efforts to update this manual.

Additional thanks to Meridian Bioscience in Cincinnati for their generous offer to make diagnostic kits available for some exercises. We also thank the following instructors for their valuable input on the revision of this manual.

Reviewers

Michelle Badon
University of Texas—Arlington

Neil Baker
Ohio State University

Christina Gan
Highline Community College

Kathryn Germain
Southwest Tennessee Community College

Sara Reed Houser
Jefferson College of Health Sciences

Terri Lindsey
Tarrant County College District - South Campus

Suzanne Schlador
North Seattle Community College

Lois Sealy
Valencia Community College

We would also like to express our appreciation to Jim Connely and Mary Jane Lampe of McGraw-Hill, Robin Reed of S4Carlisle Publishing Services, and all the others involved in publishing this manual.

Laboratory Safety

To be read by the student before beginning any laboratory work.

1. Do not eat, drink, smoke, or store food in the laboratory. Avoid all finger-to-mouth contact.
2. Never pipet by mouth because of the danger of ingesting microorganisms or toxic chemicals.
3. Wear a laboratory coat while in the laboratory. Remove it before leaving the room and store it in the laboratory until the end of the course.*
4. Wipe down the bench surface with disinfectant before and after each laboratory period.
5. Tie long hair back to prevent it from catching fire in the Bunsen burner or contaminating cultures.
6. Keep the workbench clear of any unnecessary books or other items. Do not work on top of the manual because if spills occur, it cannot be disinfected easily.
7. Be careful with the Bunsen burner. Make sure that paper, alcohol, the gas hose, and your microscope are not close to the flame.
8. All contaminated material and cultures must be placed in the proper containers for autoclaving before disposal or washing.
9. Mix cultures gently to avoid creating aerosols. Clean off the loop in a sand jar before flaming in the Bunsen burner.
10. If a culture is dropped and broken, notify the instructor. Cover the contaminated area with a paper towel and pour disinfectant over the material. After ten minutes, put the material in a broken glass container to be autoclaved.
11. Carefully follow the techniques of handling cultures as demonstrated by the instructor.
12. When the laboratory is in session, keep the doors and windows shut. Post a sign on the door indicating that a microbiology laboratory is in session.
13. Be sure you know the location of fire extinguishers, eyewash apparatus, and other safety equipment.
14. Wash your hands with soap and water after any possible contamination and at the end of the laboratory period.
15. If you are immunocompromised for any reason (including pregnancy), consult a physician before taking this class.

---

* Other protective clothing includes closed shoes, gloves (optional), and eye protection.

Laboratory Safety Agreement

To be read by the student before beginning any laboratory work.

1. Do not eat, drink, smoke, or store food in the laboratory. Avoid all finger-to-mouth contact.
2. Never pipet by mouth because of the danger of ingesting microorganisms or toxic chemicals.
3. Wear a laboratory coat while in the laboratory. Remove it before leaving the room and store it in the laboratory until the end of the course.*
4. Wipe down the bench surface with disinfectant before and after each laboratory period.
5. Tie long hair back to prevent it from catching fire in the Bunsen burner** or contaminating cultures.
6. Keep the workbench clear of any unnecessary books or other items. Do not work on top of the manual because if spills occur, it cannot be disinfected easily.
7. Be careful with the Bunsen burner.** Make sure that paper, alcohol, the gas hose, and your microscope are not close to the flame.
8. All contaminated material and cultures must be placed in the proper containers for autoclaving before disposal or washing.
9. Mix cultures gently to avoid creating aerosols. Clean off the loop in a sand jar before flaming in the Bunsen burner.
10. If a culture is dropped and broken, notify the instructor. Cover the contaminated area with a paper towel and pour disinfectant over the material. After ten minutes, put the material in a broken glass container to be autoclaved.
11. Carefully follow the techniques of handling cultures as demonstrated by the instructor.
12. When the laboratory is in session, keep the doors and windows shut. Post a sign on the door indicating that a microbiology laboratory is in session.
13. Be sure you know the location of fire extinguishers, eyewash apparatus, and other safety equipment.
14. Wash your hands with soap and water after any possible contamination and at the end of the laboratory period.
15. If you are immunocompromised for any reason (including pregnancy), consult a physician before taking this class.

---

* Other protective clothing includes closed shoes, gloves (optional), and eye protection.

** **Note to Instructor:** Modify these rules if you do not use a Bunsen burner.

I have read and understood the laboratory safety rules:

_____   _____

Signature   Date

Introduction to Microbiology

When you take a microbiology class, you have an opportunity to explore an extremely small biological world that exists unseen in our own ordinary world. Fortunately, we were born after the microscope was perfected, so we can see these extremely small organisms. At this time we also actually understand how the cell transfers the information in its genes to the operation of its cellular mechanisms so that we can study not only what bacteria do but also how they do it.

A few of these many and varied organisms are pathogens (capable of causing disease). Special techniques have been developed to isolate and identify them as well as to control or prevent their growth. The exercises in this manual will emphasize medical applications. The goal is to teach you basic techniques and concepts that will be useful to you now or can be used as a foundation for additional courses. In addition, these exercises are also designed to help you understand basic biological concepts that are the foundation for applications in all life science fields.

As you study microbiology, it is also important to appreciate the essential contributions of microorganisms as well as their ability to cause disease. Most organisms play indispensable roles in breaking down dead plant and animal material into basic substances that can be used by other growing plants and animals. Photosynthetic bacteria are an important source of the earth's supply of oxygen. Microorganisms also make major contributions in the fields of antibiotic production, food and beverage production as well as food preservation, and more recently, recombinant DNA technology. The principles and techniques demonstrated here can be applied to these fields as well as to medical technology, nursing, or patient care. This course is an introduction to the microbial world, and we hope you will find it useful and interesting.

In the next few exercises you will be introduced to several basic procedures: the use of the microscope as well as aseptic and pure culture techniques. These are skills that you will use not only throughout the course, but in any microbiology laboratory work you do in the future.

Note: The use of pathogenic organisms has been avoided whenever possible, and nonpathogens have been used to illustrate the kinds of tests and procedures that are actually carried out in clinical laboratories. In some cases, however, it is difficult to find a substitute and organisms of low pathogenicity are used. These exercises will have an additional safety precaution.

NOTES:

1

Ubiquity of Microorganisms

Getting Started

Microorganisms are everywhere—in the air, soil, and water; on plant and rock surfaces; and even in such unlikely places as Yellowstone hot springs and Antarctic ice. Millions of microorganisms are also found living within or on animals—for example, the mouth, the skin, and the intestine (all exterior to our actual tissues) support huge populations of bacteria. In fact, the interior of healthy plant and animal tissues is one of the few places free of microorganisms. In this exercise, you will sample material from the surroundings and your body to determine what organisms are present that will grow on laboratory **media.**

An important point to remember as you try to grow organisms is that there is no one condition or **medium** that will permit the growth of all microorganisms. The trypticase soy **agar** used in this exercise is a rich medium (a digest of meat and soy products, similar to a beef and vegetable broth) and will support the growth of many diverse organisms, but bacteria growing in a freshwater lake that is very low in organic compounds would find it too rich (similar to a goldfish in vegetable soup). However, organisms that are accustomed to living in our nutrient-rich throat might find the same medium lacking substances they require.

Temperature is also important. Organisms associated with warm-blooded animals usually prefer temperatures close to 37°C, which is approximately the body temperature of most animals. Soil organisms generally prefer a cooler temperature of 30°C. Organisms growing on glaciers would find room temperature (about 25°C) much too warm and would probably grow better in the refrigerator.

Microorganisms also need the correct atmosphere. Many bacteria require oxygen, while other organisms find it extremely toxic and will only grow in the absence of air. Therefore, the organisms you see growing on the plates may be only a small sample of the organisms originally present.

Definitions

Agar. A carbohydrate derived from seaweed used to solidify a liquid medium.

Colony. A visible population of microorganisms growing on a solid medium.

Inoculate. To transfer organisms to a medium to initiate growth.

Media (medium, singular). The substances used to support the growth of microorganisms.

Normal biota. The organisms usually found associated with parts of the body (previously termed *normal flora*).

Pathogen. An organism capable of causing disease.

Sterile. The absence of either viable microorganisms or viruses capable of reproduction.

Ubiquity. The existence of something everywhere at the same time.

Objectives

1. To demonstrate that organisms are **ubiquitous.**
2. To demonstrate how organisms are grown on laboratory culture media.

Reference

Nester et al. *Microbiology: A human perspective,* 6th ed. The McGraw-Hill Companies, 2009. Chapter 4.

Materials

Per team of two (or each individual, depending on number of plates available)

Trypticase soy agar (TS agar) plates, 2

Sterile swabs as needed

Sterile water (about 1 ml/tube) as needed

Waterproof marking pen or wax pencil

Procedure

First Session

1. Each pair of students should obtain two petri plates of trypticase soy agar. Notice that the lid of a petri plate fits loosely over the bottom half.
2. Label the plates with your name and date using a wax pencil or waterproof marker. Always label the bottom of the plate because sometimes you may be examining many plates at the same time and it is easy to switch the lids.
3. Divide each plate in quarters with two lines on the bottom of the petri plate. Label one plate 37°C and the other 25°C (figure 1.1).
4. **Inoculate** the 37°C plate with samples from your body. For example, moisten a sterile swab with sterile water and rub it on your skin and then on one of the quadrants (be careful not to gouge the agar). Try touching your fingers to the agar before and after washing, or place a hair on the plate. Try whatever interests you. (Be sure to place all used swabs into an autoclave container or bucket of disinfectant after use.)
5. Inoculate the plate labeled 25°C (room temperature) with samples from the room. It is easier to pick up a sample if the swab is moistened in sterile water first. **Sterile** water is water in which there are no living organisms, and is used so that your results will not be contaminated. Try sampling the bottom of your shoe or some dust, or press a coin or other objects lightly on the agar. Be sure to label each quadrant so that you will know what you used as inoculum.
6. Incubate the plates at the temperature written on the plate. Place the plates in the incubator or basket upside down. This is important because it prevents condensation from forming on the lid and dripping on the agar below. The added moisture would permit colonies of bacteria to run together.

Second Session

Handle all plates with colonies as if they were potential **pathogens,** disease-causing organisms. Follow your instructor's directions carefully.

Figure 1.1 Plates labeled on the bottom for ubiquity exercise.

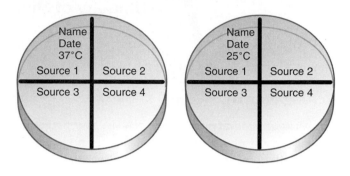

Note: For best results, the plates incubated at 37°C should be observed after 2 days, but the plates at room temperature will be more interesting at about 5–7 days. If possible, place the 37°C plates either in the refrigerator or at room temperature after 2 days so that all the plates can be observed at the same time.

1. Examine the plates you prepared in the first session and record your observations on the report sheet for this exercise. There will be basically two kinds of **colonies:** fungi (molds) and bacteria. Mold colonies are usually large and fluffy, the type found on spoiled bread. Bacterial colonies are usually soft and glistening and tend to be cream colored or yellow. Compare your colonies with color plates 1 and 2.
2. When describing the colonies include:
   a. relative size as compared to other colonies
   b. shape (round or irregular)
   c. color
   d. surface (shiny or dull)
   e. consistency (dry, moist, or mucoid)
   f. elevation (flat, craterlike, or conical)
3. There may be surprising results. If you pressed your fingers to the agar before and after washing, you may find more organisms on the plate after you washed your hands. The explanation is that your skin has **normal biota** (organisms that are always found growing on your skin). When you wash your hands, you wash off the organisms you have picked up from your surroundings as well as a few layers of skin. This exposes more of your normal biota; therefore, you may see different

colonies of bacteria before you wash your hands than afterward. Your biota is important in preventing undesirable organisms from growing on your skin. Hand washing is an excellent method for removing pathogens that are not part of your normal biota.

4. (Optional) If desired, use these plates to practice making simple stains or Gram stains as described in exercises 5 and 6.

Note: In some labs, plates with molds are opened as little as possible and immediately discarded in an autoclave container to prevent contaminating the lab with mold spores.

5. Follow the instructor's directions for discarding plates. All agar plates are autoclaved before washing or discarding in the municipal garbage system.

NOTES:

EXERCISE

1

Laboratory Report: Ubiquity of Microorganisms

Results

Room Temperature (about 25°C) Plate				
	Plate Quadrant			
	1	2	3	4
Source				
Colony appearance				

37°C Plate				
	Plate Quadrant			
	1	2	3	4
Source				
Colony appearance				

Questions

1. Give three reasons why all the organisms you placed on the trypticase soy agar plates might not grow.

2. Why were some agar plates incubated at 37°C and others at room temperature?

3. Why do you invert agar plates when placing them in the incubator?

4. Name one place that might be free of microorganisms.

EXERCISE

2

Pure Culture and Aseptic Technique

Getting Started

Aseptic Technique and Streak Plate Technique

The two goals of **aseptic** (or **sterile**) technique are to prevent contamination of your culture with organisms from the environment and to prevent the culture from contaminating you or others.

In this exercise, you will transfer sterile broth back and forth from one tube to another using the aseptic technique. The goal is to do it in such a way that no organism in the environment enters the tubes. You will be using both a sterile pipet and a flamed loop on the same set of broths. After you have practiced aseptically transferring the broth, you will **incubate** the broth tubes for a few days to determine if they are still sterile. If you used good technique, the broth will still be clear; if organisms were able to enter from the environment, the broth will be cloudy from the bacterial growth.

When you can successfully transfer sterile broth aseptically, you can use the same technique to transfer a pure culture without contaminating it or the environment.

The streak plate is another essential technique. It permits you to isolate a colony formed by a single cell from a mixture containing millions of cells. You will start with a broth culture containing two different organisms and separate them into two different colony types. By spreading the bacteria around on the surface of the plate, the cells are able to form individual colonies separated from other individual colonies. This technique is used to obtain a **pure culture.**

Definitions

Aseptic. Free of contamination.

Incubate. Store cultures under conditions suitable for growth, often in an incubator.

Pure culture. A population of cells resulting from the growth of a single cell.

Sterile. Free of viable bacteria or viruses.

Objectives

1. To learn aseptic technique procedures and their importance.
2. To learn to isolate colonies using the streak plate technique.

Reference

Nester et al. *Microbiology: A human perspective*, 6th ed. The McGraw-Hill Companies, 2009. Chapter 4, Section 4.3.

Aseptic Technique

Broth-to-Broth Transfer with a Wire Loop

Materials

Per student
   Tubes of TS broth, 3
   Inoculating loop
   Broth culture of *Escherichia coli*

Procedure

1. Always label tubes before adding anything to them. In this exercise, you will be transferring sterile broth from one tube to another, so two tubes can be labeled "P," for practice. Label the third tube "C," for control. Always include the date on anything you label.
2. Grip the loop as you would a pencil and flame the wire portion red hot. Hold it at an angle so that you do not burn your hand (figure 2.1).
3. After the loop has cooled for a few seconds, pick up a practice tube in the other hand and remove the cap of the tube with the little finger (or the fourth and little fingers) of the hand holding the loop.
4. Flame the mouth of the tube by passing it through a Bunsen burner flame, and then use

Figure 2.1 Aseptic technique for removing a loopful of broth culture. *(a)* Hold the culture tube in your left hand and the loop in your right hand (reverse if you are left-handed). Flame the loop to sterilize it. *(b)* Remove the culture tube plug or cap, and flame the mouth of the culture tube. *(c)* Insert the sterile loop into the culture tube. *(d)* Remove the loopful of inoculum, and flame the mouth of the culture tube again. *(e)* Replace the culture tube plug or cap. Place the culture tube in a test tube rack. *(f)* Reflame the loop.

check to be sure that your loop is a complete circle without a gap.

5. Set down the first tube and pick up the second practice tube. Remove the cap, flame it, and deposit a loopful of material into the liquid of the second tube. Withdraw the loop, flame the tube, and then replace the cap. Be sure to flame the loop before setting it on the bench (your loop would normally be contaminated with the bacteria you were inoculating). It is usually convenient to rest the hot loop on the edge of the Bunsen burner.

6. Using the loop as before, inoculate the control tube with the broth culture provided by the instructor. Incubate it with the other tubes.

7. When learning aseptic technique, it is better to hold one tube at a time; later, you will be able to hold two or three tubes at the same time.

Note: Many laboratories no longer flame the mouth of the tube, especially if the tubes are plastic. However some laboratories still follow this procedure.

Broth-to-Broth Transfer with a Pipet

Note: Sterile pipets are used when it is necessary to transfer known amounts of material. Some laboratories use plastic disposable pipets and others use reusable glass pipets. Be sure to follow the instructor's directions for proper disposal after use (never put a used pipet on your bench top). Mouth pipetting is dangerous and is not permitted. A variety of bulbs or devices are used to draw the liquid up into the pipet, and your laboratory instructor will demonstrate their use (figure 2.2).

The same broth tubes used for practice with the loop can be used to practice pipetting the broth back and forth.

> ### Materials
>
> TS broth tubes from previous procedure
> 1-ml pipet
> Bulb or other device to fit on end of the pipet

the sterile loop to obtain a loopful of liquid from the tube. Flame the mouth of the tube again and replace the cap. If you have trouble picking up a loopful of material,

Figure 2.2 Various devices for filling pipets.

Small rubber bulb

(a)

Bulb and valve pipet filler

(b)

Pi-Pump® pipettors
Blue—for pipets up to 2 ml
Green—for pipets up to 10 ml

(c)

Automatic pipet aid

(d)

Procedure

First Session

1. Open a sterile pipet at the top, insert a bulb on the end, then carefully remove the pipet from the package or canister without touching the tip. Grip the pipet as you would a pencil. The pipet is plugged with cotton to filter the air going into it. Discard the pipet if liquid inadvertently wets the plug—air will no longer enter the pipet and the measured liquid will not flow out. Notify your instructor if the bulb is contaminated.

2. Pick up a tube with your other hand and remove the cap with the little finger of the hand holding the pipet. Flame the tube. Expel air from the rubber bulb and insert the pipet tip into the liquid. Note that the liquid must be drawn to the 0 mark for 1 ml when using a 1-ml pipet. Draw the liquid up to the desired amount, remove the pipet, flame the tube, replace the cap, and then put the tube back in the rack.

3. Pick up the second tube and repeat the steps used with the first tube except that the liquid is expelled into the tube.

4. Repeat the above steps with the same tubes until you feel comfortable with the procedure.
5. Dispose of the pipet as directed.
6. Incubate the tubes at 37°C until the next period.

Second Session

1. Observe the control broth and note if it is cloudy. This turbidity indicates growth and demonstrates that the broth can support the growth of organisms.
2. Observe the practice tubes of broth for turbidity. Compare them to an uninoculated tube of broth. If they are cloudy, organisms contaminated the broth during your practice and grew during incubation. With a little more practice, you will have better technique. If the broths are clear, there was probably no contamination and you transferred the broth without permitting the entry of any organisms into the tubes.
3. Record results.

Streak Plate Technique

Materials

Per student

Trypticase soy agar (TS agar) plates, 2

Broth culture containing a mixture of two organisms, such as *Micrococcus* and *Staphylococcus*

Procedure

First Session

1. Label the agar plate on the bottom with your name and date.
2. Divide the plate into three sections with a *T* as diagrammed (figure 2.3).
3. Sterilize the loop in the flame by heating the whole length of the wire red hot. Hold it at an angle so you do not heat the handle or roast your hand.
4. Gently shake the culture to be sure both organisms are suspended. Aseptically remove

Figure 2.3 (a–g) Preparation of a streak plate.

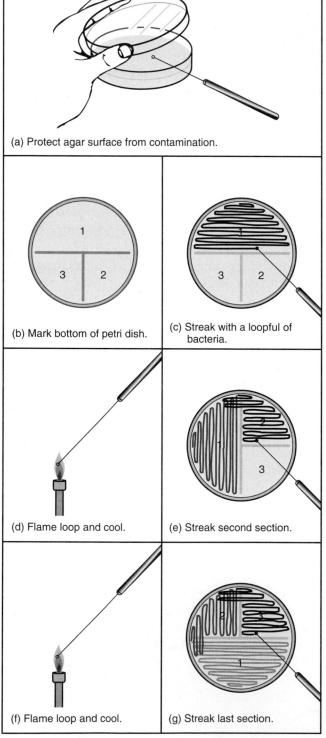

(a) Protect agar surface from contamination.

(b) Mark bottom of petri dish.

(c) Streak with a loopful of bacteria.

(d) Flame loop and cool.

(e) Streak second section.

(f) Flame loop and cool.

(g) Streak last section.

a loopful of the culture and, holding the loop as you would a pencil, spread the bacteria on section 1 of the plate by streaking back and forth. The more streaks, the better chance of an isolated colony. As you work, partially cover the petri dish with the cover to minimize organisms falling on the plate from the air. Use a gliding motion and avoid digging into the agar. Don't press the loop into the surface. If your loop is not smooth or does not form a complete circle, it can gouge the agar and colonies will run together. Note that you can see the streak marks if you look carefully at the surface of the plate.

5. Burn off all the bacteria from your loop by heating it red hot. This is very important because it eliminates the bacteria on your loop. Wait a few seconds to be sure the loop is cool.

6. Without going into the broth again, streak section 2 (see figure 2.3) of the petri plate. Go into section 1 with about three streaks and spread by filling section 2 with closely spaced streaks. Be very careful not to touch the streaks in section 1 again.

7. Again heat the loop red hot. Go into section 2 with about three streaks and spread by filling section 3 with streaks. The more streaks you are able to make, the greater will be your chance of obtaining isolated colonies.

8. Heat loop red hot before placing it on the bench top. Usually you can rest it on some part of the Bunsen burner so that it can cool without burning anything.

9. Repeat the procedure with a second plate for additional practice.

10. Incubate the plates, inverted, in the 37°C incubator.

Second Session

Observe your streak plates and record results.

EXERCISE

2

Laboratory Report: Pure Culture and Aseptic Technique

Results

	Clear	Turbid
Tube 1		
Tube 2		
Control		

	Number of Colonies Isolated
Streak plate 1	
Streak plate 2	

Did you obtain isolated colonies of each culture?

Questions

1. What is the definition of a pure culture?

2. Why is sterile technique important? Give two reasons.

3. What did the control broth inoculated with *Escherichia coli* demonstrate?

4. What is the purpose of a streak plate?

5. Why is it important to avoid digging into the agar with the loop?

6. Is there anything you can do to improve your streak plate technique?

EXERCISE

3

Introduction to the Compound Light Microscope

Getting Started

Microbiology is the study of organisms too small to be seen with the naked eye, including a vast array of bacteria, viruses, protozoa, fungi, and algae. van Leeuwenhoek, a Dutch merchant in the late 17th century whose hobby was lens making, was the first to see these previously unknown creatures. His microscope consisted of one simple lens, but it was enough to observe some of these tiny living things (figure 3.1). Although he made drawings of

some of these organisms, he did not suspect that they were essential for the existence of our world or that a small percentage was responsible for contagious diseases. It was only after the development of the compound microscope almost 200 years later that it was possible to understand the role of microorganisms in disease. After the microscope was perfected in the 1870s and 1880s, real progress was made in determining the actual cause of disease (see Appendix 7 for a history of the development of the microscope). In 1877, Robert Koch saw bacteria

Figure 3.1 Model of a van Leeuwenhoek microscope. The original was made in 1673 and could magnify the object being viewed almost 300 times. The object being viewed is brought into focus with the adjusting screws. This replica was made according to the directions given in the *American Biology Teacher* 30:537, 1958. Note its small size. Photograph Courtesy of J.P. Dalmasso

Lens

Object being viewed

Adjusting screws

Viewing side

1 inch

in the blood from an animal with anthrax, and for the first time it could be seen that bacteria, not swamp gas or evil spirits, could cause disease.

The modern **compound light microscope** consists of two lens systems. The first is the objective lens, which is closest to the material on the slide, and the second is the eyepiece, or ocular lens, which magnifies the image formed by the objective lens. The total **magnification** is found by multiplying the eyepiece lens magnification by the objective lens magnification. For example, if the eyepiece lens magnification is 10X and the objective lens magnification is 45X, the total magnification is 450 diameters.

Although it is possible to put additional magnifying glasses on top of the eyepiece lenses of the microscope, they would not improve the ability to see more detail. The reason is that the actual limiting factor of the light microscope is the **resolution**. This is the ability to distinguish two close objects as distinct from one another rather than as one round hazy object. The resolution of a lens is limited by two factors: the angle of the lens, and the wavelength of light entering the microscope. When using an objective lens of magnification 100X and light is optimal, a compound light microscope has a maximum magnification of about 1,000X.

A magnification of 1,000X is sufficient to easily visualize single-celled organisms such as algae and fungi. Bacteria, however, still appear very small (about the size of the letter *l* on a printed page), and their appendages, such as flagella, can not be observed at all. Viruses are also too small to be seen in a light microscope.

To surmount the limitations of light and lens, the electron microscope, which uses electrons instead of light, was developed in the 1930s. It can magnify objects 100,000X and thus permits the visualization of viruses and subcellular bodies. Because the electron microscope is a very large piece of equipment requiring specialized techniques, it usually is found only in universities or research facilities.

In this exercise you will have an opportunity to become familiar with a compound light microscope and learn how to use and care for it. You will also prepare **wet mounts** and learn to observe previously stained and unstained organisms. The microscope is an expensive and complex piece of equipment. Treat it with great care.

The Parts of the Microscope

The eyepieces are the lenses at the top of the microscope. They usually have a magnification of 10X. (see figure 3.2.)

Figure 3.2 Modern bright-field compound microscope. Courtesy of Carl Zeiss, Inc.

Eyepiece (Ocular): a magnifying lens, usually about 10X

Coarse-adjustment focusing knob

Fine-adjustment focusing knob

Objective nosepiece and attached objective lenses, each with a different magnification. The total magnification equals the product of the objective lens employed with the ocular lens.

Specimen stage: the platform on which the slide is placed

Iris diaphragm lever: regulates the amount of light that enters the objective lens

Condenser: focuses the light

Base with illuminating light source

Most microscopes have at least three objective lenses. They usually are:

Low power	10X
High power	40X
Oil immersion	100X

These lenses can be rotated and the desired lens clicked into place.

The stage is below the objective lens. It usually has a device called a mechanical stage for holding the slide as well as knobs that permit the slide to be moved smoothly while viewing.

Below the stage is the **condenser**, which focuses the light on the slide. If it is lowered, the amount of light is reduced, but the resolution is also lowered. For our purposes the condenser should remain at its highest position under the stage.

A lever on top of the condenser but under the stage controls the **iris diaphragm**. The diaphragm is important for adjusting the amount of light illuminating the slide. The higher the magnification, the greater the amount of light that is needed.

The light source is at the base of most microscopes. Usually the light source is set at maximum and the amount of light on the slide is adjusted with the iris diaphragm.

Precautions for Proper Use and Care of the Microscope

1. Carry the microscope with both hands. Keep it upright. If the microscope is inverted, the eyepieces may fall out.
2. Do not remove the objective or ocular lenses for any purpose.
3. If something seems stuck or you have problems making adjustments, do not apply force. Consult the instructor.
4. Never touch or wipe the lenses with anything but lens paper. Clean the lens by gently drawing a flat piece of lens paper across it. The presence of foreign particles can be determined by rotating the eyepiece lenses manually as you look through the microscope. A pattern that rotates is evidence of dirt. If wiping the lenses with lens paper does not remove the dirt, consult the instructor. It may be on the inside surface of the lens.
5. Before storing the microscope:

    If the microscope has an adjustable tube, rack it down so that the microscope can be stored more easily.
    Make certain the eyepiece lenses are clean. Sweat deposits from your eyes can etch the glass.
    IMPORTANT: If you have used the oil immersion lens be sure to wipe off the oil. If not removed it can leak into the lens and cause severe damage.

Definitions

Compound microscope. A microscope with more than one lens system.

Condenser. A structure located below the microscope stage that contains a lens for focusing light on the specimen as well as an iris diaphragm.

Iris diaphragm. An adjustable opening that regulates the amount of light illuminating the specimen.

Magnification. The microscope's ability to optically increase the specimen size.

Resolution. The smallest separation that two structural forms—two adjacent cilia for example—must have in order to be distinguished optically as separate.

Wet mount. A laboratory technique wherein a microscopic specimen is placed on the surface of a slide and covered with a coverslip.

Objectives

1. To introduce the parts of the microscope.
2. To explain the use and proper care of the microscope.
3. To examine stained and unstained material.

References

Gerhardt, Phillip, ed. *Manual for general and molecular bacteriology.* Washington, D.C.: American Society for Microbiology, 1993, paperback.

Nester et al. *Microbiology: A human perspective,* 6th ed. The McGraw-Hill Companies, 2009. Chapter 3, Section 3.1.

Materials

Prepared slides of a printed letter *e* with coverslips (optional)

Prepared stained slides of protozoa and other large cells

Several jars of hay infusion containing protozoa and algae

Yeast

Procedure

Important note: In this introduction to the microscope, the 100X oil objective lens (usually labeled with a broad band) will not be used. It is the most expensive lens and its use with immersion oil will be explained in exercise 4. Be particularly careful not to hit this lens on the stage.

1. Place the microscope on a clear space on your bench, away from any flame or heat source. Identify the different parts with the aid of figure 3.2 and the Getting Started section "The Parts of the Microscope."
2. Before using the microscope be sure to read the Getting Started section "Precautions for Proper Use and Care of the Microscope."
3. Obtain either a prepared slide with the printed letter *e* covered with a coverslip or a large stained specimen of protozoa, fungi, or algae. Place the slide in the mechanical stage, coverslip side up, and turn on the light at the base of the microscope. Move the eyepiece lenses apart, and then while looking through the lenses, push them together until you see one circle. This is the ocular width of your eyes. If you note the number on the scale between the lenses, you can set the microscope to this number each time you use it.
4. Rotate the low-power objective lens (10X) into place. When looking from the side, bring the lens as close to the slide as possible. Then when looking through the microscope, use the large knob, called the course-adjustment knob, to raise the lens until the object is in focus. Increase and decrease the amount of light with the iris diaphragm lever to determine the optimal amount of light. Continue to focus using the fine-adjustment

knob, which is the smaller knob. Remember that the objective lens should never touch the surface of the slide or coverslip.
5. Move the slide back and forth. When viewing objects through the microscope, the image moves in the opposite direction than the slide is actually moving. It takes awhile to become accustomed to this phenomenon, but later it feels normal.
6. View the specimen at a higher magnification. Rotate the 40X objective lens into place (sometimes called the high-dry objective lens). Be very careful not to hit the slide or the stage with the oil immersion lens. Notice the change in the amount of light needed. Also note that the specimen is almost in focus. Most microscopes are parfocal, meaning that the objective lens can be rotated to another lens and the slide remains in focus. View several prepared slides until you become comfortable with the microscope. Always start focusing with the low-power objective lens, then move to the high-dry objective lens.

Wet Mounts

1. Prepare a wet mount (figure 3.3). Clean a glass slide with a mild cleansing powder such as Bon Ami or as directed by your instructor.
2. Place a drop (about the size of a small pea) of hay infusion or other material on the slide using a dropper. Carefully place a coverslip on edge next to the drop and slowly lower it so that it covers the drop. Try to avoid air bubbles. If the drop is so big it leaks out from under the coverslip, discard the slide into a container designated by the instructor and prepare another wet mount.
3. Examine the wet mount using the low-power objective lens, and then switch to the high-power objective lens. When viewing

Figure 3.3 Preparation of a wet mount.

unstained material, it is necessary to reduce the amount of light to increase the contrast between the cells and the liquid. If you are having difficulty focusing on the material, try to focus on the edge of the coverslip and then move the slide into view.

4. Draw examples of the material viewed. Indicate the total magnification.

5. Dispose of your slides as directed by your instructor. The material on the slide is still viable, so the slides should be boiled or autoclaved.

6. Wipe the lenses with lens paper. Return your microscope carefully to its storage space.

NOTES:

Name _____ Date _____ Section _____

EXERCISE

3

Laboratory Report: Introduction to the Compound Light Microscope

Results

1. Draw four fields of the material you observed. Indicate which objective lens was used and the total magnification. If possible show the same field or material at two different magnifications.

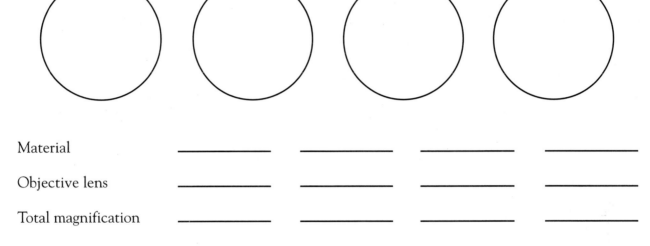

Material   _____   _____   _____   _____

Objective lens   _____   _____   _____   _____

Total magnification   _____   _____   _____   _____

2. On your microscope, what is the magnification of:

    a. the eyepiece lenses?
    b. the low-power lens?
    c. the high-dry lens?

Questions

1. When you increase the magnification, is it necessary to increase or decrease the amount of light?

2. When looking at unstained material, do you need more or less light than what is needed to view a stained preparation?

3. Why is it convenient to have a parfocal microscope?

4. Why couldn't you see a virus with your microscope even if you increased the eyepiece lens magnification to 100X?

5. Why was Koch's observation of bacteria in blood so significant?

6. What is the only material that should be used to wipe the lens?

EXERCISE 4

The Use of the Immersion Lens and Determining Motility in a Wet Mount

Getting Started

Some bacteria can move through liquid using flagella (flagellum, singular). These flagella are much smaller than those seen on protozoa and can only be visualized in the light microscope if special stains are used.

Although bacterial cell motility is usually determined by the semisolid agar stab inoculation method, it is sometimes determined by direct microscopic examination of wet mounts. This has the advantage of immediate results and also gives some additional information about the shape and size of the organism. Sometimes it is possible to see how the cells are arranged, as in tetrads or chains. However, these groupings are easily broken up in the process of making the wet mount so the designations are not reliable.

There are also disadvantages for determining motility from a wet mount. Some types of media inhibit motility and cultures not actively multiplying may no longer be motile.

When observing bacteria in wet mounts it is important to distinguish true motility from other kinds of motion. Evaporation at the edge of the coverslip causes convection currents to form. The suspended cells appear to be moving along in a stream flowing to the edge of the coverslip. If cells are truly motile, they swim in random directions.

"Brownian movement" is a form of movement caused by molecules in liquid striking a solid object—in this case, a bacterial cell—causing the cell to slightly bounce back and forth. Bacterial cells appear to jiggle in liquid, but if the cell is actually motile it moves from one point to another.

Using the Oil Immersion Lens

To observe bacteria, either stained or in a wet mount, it is necessary to use the oil immersion lens. This lens (100X) permits you to attain the highest magnification of the light microscope. It is the longest and usually has a band around it. Extra care must be taken when using it because not only is it the most valuable, but because of its length it is possible to

hit and break the slide when attempting to focus with it. Also care must be taken not to hit the lens on the stage when rotating a lens into place.

The reason this lens is called the oil immersion lens is that **immersion oil** is put on the top of a stained slide or wet mount and the lens is then carefully lowered into the oil. The purpose of the oil is to prevent **refraction** of the light entering the lens. Light bends each time it goes through a different medium. (This can be seen in the appearance of a spoon in a glass of water.) To prevent the light's bending from its source in the microscope through air and then through the glass lens, the **refractive index** of the oil is the same as glass (figure 4.1). This can be observed if your dispensing bottle for oil has a glass dropper. The dropper will disappear when you immerse it in the oil because the light does not bend between the immersion oil and the glass (see Appendix 7 for additional information).

In this exercise the immersion lens will be used to observe the motility of bacteria.

Definitions

Refraction. The bending of light as it passes from one medium to another.

Immersion oil. Oil placed on a slide to minimize refraction of the light entering the lens.

Figure 4.1 Oil immersion lens.

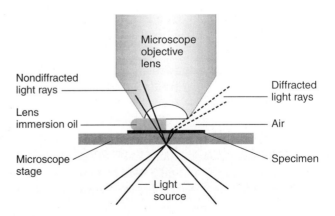

Refractive index. The ratio of the velocity of light in the first of two media to its velocity in the second medium as it passes from one medium into another medium.

Objectives

1. To learn to use the oil immersion lens.
2. To observe motile bacteria in a wet mount preparation.

Reference

Nester et al. *Microbiology: A human perspective,* 6th ed. The McGraw-Hill Companies, 2009. Chapter 3, Section 3.2.

Materials

Cultures of the following:

Overnight cultures in nutrient broth of the following:

   Spirillum volutans

   Bacillus cereus

   Pseudomonas

Overnight culture in TS broth of the following:

   Staphylococcus epidermidis

Prepared stained slides

Vaseline (optional)

Toothpicks (optional)

Procedure

Use of the Oil Immersion Lens

1. Place a stained slide in your mechanical stage and focus on the specimen, first with the fine adjustment knob, then with the course adjustment knob. (see Exersise 3.) Increase or decrease the light with the iris diaphragm for optimal viewing.
2. Swing the lens away from the slide and place a drop of immersion oil directly on top of the slide.
3. Carefully rotate the oil immersion lens into the oil and click into place. Focus the microscope using the fine adjustment knob. If the microscope is parfocal, it should take very little adjustment. Notice that you must increase the amount of light as you increase the magnification.

4. After you feel comfortable using the oil immersion lens, swing it away from the slide, remove the slide, and take the oil off the slide with a piece of lens paper or as directed by your instructor.
5. Also remove the oil from the lens with lens paper if you are not going to continue.

Determining Motility with a Wet Mount

1. Using aseptic technique, sterilize a loop in the Bunsen burner, permit it to cool, and place a drop of culture on a clean slide. If the drop isn't the size of a small pea, add another (this takes a little practice). Remember to sterilize the loop before putting it back in the culture. It is easier to use a separate slide for each culture when beginning.
2. Carefully place a coverslip on one edge next to the drop and lower it down. (see figure 3.3.) If some of the drop seeps out from under the coverslip, it may contaminate the lens. Dispose of the slide as directed by the instructor—it cannot be simply washed off because it contains viable microorganisms.
3. Focus the slide as usual, first with the low-power lens, then with the high-dry lens. You will have to greatly reduce the amount of light using the iris diaphragm because there is very little contrast between the broth and the unstained organisms.
4. Place a drop of immersion oil on the coverslip of the wet mount and carefully click the oil immersion lens into place. Refocus the microscope, being very careful not to touch the slide with the lens. You should only need to use the fine-adjustment knob.
5. Observe the bacteria to determine if they are motile. Distinguish among Brownian movement, convection currents, and true motility (see the Getting Started Section).
6. If you want to have more time to observe your wet mounts, you can coat the edge of the coverslip with vaseline using a toothpick before covering the drop on the slide. This will prevent the specimen from evaporating as quickly.
7. Dispose of your slides as directed by your instructor.
8. Record your results.

EXERCISE

4

Laboratory Report: The Use of the Immersion Lens and Determining Motility in a Wet Mount

Results

Questions

1. Draw a picture of each culture as observed under the oil immersion lens.

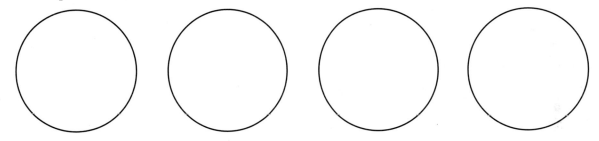

   Which of the cultures are motile?

2. When you are using the oil immersion lens, what is the total magnification?

3. How do cells appear when Brownian movement is causing their motion?

4. How do cells appear when convection currents are causing their motion?

5. What must you do without fail before storing your microscope?

NOTES:

*I*NTRODUCTION to Staining of Microorganisms

Bacteria are difficult to observe in a broth or wet mount because there is very little contrast between them and the liquid in which they are suspended. This problem is solved by staining bacteria with dyes. Although staining kills bacteria so their motility cannot be observed, the stained organisms contrast with the surrounding background and are much easier to see. The determination of the shape, size, and arrangement of the cells after dividing is useful in identifying an organism. These can be demonstrated best by making a smear on a glass slide from the clinical material, a broth culture, or a colony from a plate, then staining the smear with a suitable dye. Examining a stained preparation is one of the first steps in identifying an organism.

Staining procedures used here can be classified into two types: the simple stain and the multiple stain. In the **simple stain,** a single stain such as methylene blue or crystal violet is used to dye the bacteria. The most commonly used simple stains are basic dyes. These dyes are positively charged, making it possible to stain the negatively charged bacteria. The shape and the grouping of the organisms can be determined, but all organisms (for the most part) are stained the same color. Another kind of simple stain is the negative stain. In this procedure, the organisms are mixed with a dye and permitted to dry. When they are observed, the organisms are clear against a dark background.

The multiple stain involves more than one stain. The best known example is the differential Gram stain, which is widely used. After staining, some organisms appear purple and others pink, depending on the structure of their cell wall.

Special stains are used to visualize specific structures of bacteria. In contrast with eukaryotic organisms, prokaryotic organisms have relatively few morphological differences. Several of these structures, such as endospores, capsules, acid-fast cell walls, storage bodies, and flagella, can be seen with special stains. In the next two exercises, you will have an opportunity to stain bacteria with a variety of staining procedures and observe these structures.

NOTES:

EXERCISE 5

Simple Stains: Positive and Negative Stains

Getting Started

Two kinds of single stains will be done in this exercise: the simple stain and the negative stain. Microbiologists most frequently stain organisms with the Gram stain, but in this exercise a simple stain will be used to give you practice staining and observing bacteria before doing the more complicated multiple, or **differential, stains.**

After you have stained your bacterial **smears,** you can examine them with the oil immersion lens, which will allow you to distinguish the morphology of different organisms. The typical bacteria you will see are about 0.5–1.0 **micrometer** (μm) in width to about 2–7 μm long and are usually rods, cocci, or spiral-shaped. Sometimes rods are referred to as bacilli, but since that term is also a genus name (*Bacillus)* for a particular organism, the term rod is preferred.

Another kind of simple stain is the **negative stain.** Although it is not used very often, it is advantageous in some situations. Organisms are mixed in a drop of nigrosin or India ink on a glass slide. After drying, the organisms can be observed under the microscope as clear areas in a black background. This technique is sometimes used to observe **capsules** or **inclusion bodies.** It also prevents eyestrain when many fields must be scanned. The dye tends to shrink away from the organisms, causing cells to appear larger than they really are.

In both of these simple stains, you will be able to determine the shape of the bacteria and the characteristic grouping after cell division (as you did in the wet mounts). Some organisms tend to stick together after dividing and form chains or irregular clumps. Others are usually observed as individual cells. However, this particular characteristic depends somewhat on how the organisms are grown. *Streptococcus* bacteria form long, fragile chains in broth, but if they grow in a colony on a plate, it is sometimes difficult to make a smear with these chains intact.

Definitions

Capsule. A gelatinous material coating the outside of the cell.

Differential stain. A procedure that stains specific morphological structures—usually a multiple stain.

Inclusion bodies. Granules of storage material such as sulfur that accumulate within some bacterial cells.

Micrometer. (abbreviated μm) The metric unit used to measure bacteria. It is 10^{-6} m (meter) and 10^{-3} mm (millimeter).

Negative stain. A simple stain in which the organisms appear clear against a dark background.

Simple stain. A procedure for staining bacteria consisting of a single stain.

Smear. A dried mixture of bacteria and water (or broth) placed on a glass slide in preparation for staining.

Objectives

1. To learn to prepare and stain a bacterial smear using a simple stain.
2. To observe stained organisms under the oil immersion lens.
3. To prepare and observe a negative stain.
4. To observe the various morphologies and arrangements of bacteria in stained preparations.

References

Gerhardt, Philip, ed. *Manual for general and molecular bacteriology.* Washington, D.C.: American Society for Microbiology, 1993, paperback.

Nester et al. *Microbiology: A human perspective,* 6th ed. The McGraw-Hill Companies, 2009. Chapter 3, Section 3.2.

Materials

Cultures of the following:

  Bacillus subtilis or *Bacillus cereus*

  Staphylococcus epidermidis

  Streptococcus mutans

  Micrococcus luteus

Staining bottles with

  crystal violet

  methylene blue

  safranin

Glass slides

Wax pencil or waterproof marking pen

Tap water in small dropper bottle (optional)

Inoculating loop

Alcohol sand bottle (a small screw cap bottle half full of sand and about three-quarters full of 95% alcohol) (figure 5.1)

Disposable gloves

Procedure

Simple Stain

1. Clean a glass slide by rubbing it with slightly moistened cleansing powder such as Boraxo or Bon Ami. Rinse well and dry with a paper towel. New slides should be washed because sometimes they are covered with a protective coating.

2. Draw two or three circles with a waterproof pen or wax pencil on the underside of the slide. If

Figure 5.1 Alcohol sand bottle and inoculating loop.

Sand bottle

Inoculating loop

Alcohol

Sand

Figure 5.2 Slide with three drops of water. Three different bacteria can be stained on one slide.

Frosted portion

Water

the slide has a frosted portion, you can also write on it with a pencil. This is useful because it is easy to forget the order in which you placed the organisms on the slide and you can list them, for instance, from left to right (figure 5.2).

3. Add a drop of water to the slide on top of each of the circles. Use your loop to transfer tap water or use water from a dropper bottle. This water does not need to be sterile. Although there are some organisms (nonpathogens) in municipal water systems, there are too few to be seen.

   If you are preparing a smear from a broth culture, as you will do in the future, add only the broth to the slide. Broth cultures are relatively dilute, so no additional water is added.

4. Sterilize a loop by holding it at an angle in the flame of the Bunsen burner. Heat the entire wire red hot, but avoid putting your hand directly over the flame or heating the handle itself (figure 5.3).

5. Allow the loop to cool a few seconds, then remove a small amount of a bacterial culture and suspend it in one of the drops of water on the slide (see figure 5.3). Continue to mix in bacteria until the drop becomes slightly turbid (cloudy). If your preparation is too thick, it will stain unevenly, and if it is too thin, you will have a difficult time finding organisms under the microscope. In the beginning, it may be better to err on the side of having a slightly too turbid preparation—at least you will be able to see organisms and you will learn from experience how dense to make the suspension.

6. Heat the loop red hot again. It is important to burn off the remaining organisms so that you will not contaminate your bench top. If you then rest your loop on the side of your Bunsen burner, it can cool without burning anything on the bench.

   Sometimes the cell material remaining on the loop spatters when heated. To prevent

Figure 5.3 Preparation of a bacterial smear.

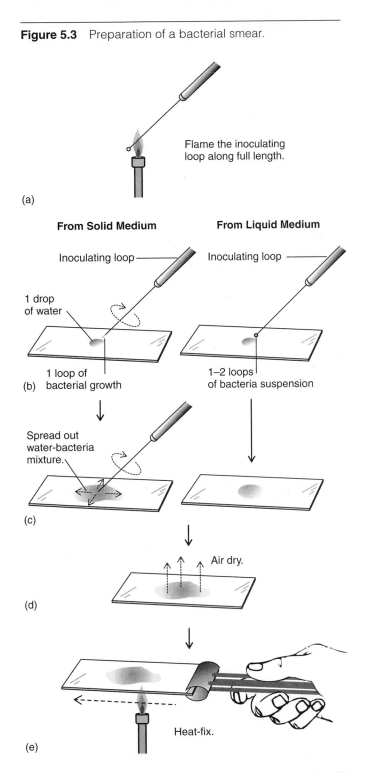

Flame the inoculating loop along full length.

(a)

From Solid Medium   **From Liquid Medium**

Inoculating loop ——   Inoculating loop ——

1 drop of water

1 loop of bacterial growth   1–2 loops of bacteria suspension

(b)

Spread out water-bacteria mixture.

(c)

Air dry.

(d)

Heat-fix.

(e)

become distorted. Place the slide off to the side of the bench so that you can proceed with other work.

8. When the slide is dry (in about 5–10 minutes), heat-fix the organisms to the slide by quickly passing it through a Bunsen burner flame two or three times so that the bottom of the slide is barely warm. This step causes the cells to adhere to the glass so they will not wash off in the staining process (figure 5.3).

9. Place the slide on a staining loop over a sink or pan. Alternatively, hold the slide over the sink with a forceps or clothespin. Cover the specimen with a stain of your choice—crystal violet is probably the easiest to see (figure 5.4).

10. After about 20 seconds, pour off the stain and rinse with tap water (figure 5.4).

11. Carefully blot the smear dry with a paper towel. Do not rub the slide from side to side as this will remove the organisms. Be sure the slide is completely dry (figure 5.4).

12. Observe the slide under the microscope. Since you are looking at bacteria, you must use the oil immersion lens in order to see them. Focus the slide on low power, then cover the smear with immersion oil and move the immersion lens into place. If your microscope is **parfocal,** it should be very close to being in focus. Note that no coverslip is used when looking at stained organisms.

    Never move the immersion lens toward the slide while looking through the microscope. You may hit the slide with the lens and damage the lens. When you have a particularly thin smear, it is sometimes helpful to put a mark on the slide near the stain using a marking pen. It is easy to focus on the pen mark, and you will know that you have the top of the slide in focus and can then search for the smear.

13. Record your results.

14. You can save your stained slide with the oil on it. If you do not want to save the slide, simply clean it with cleanser and water. The staining procedure kills the bacteria, so the slide does not need to be boiled before cleaning.

this, some laboratories remove bacterial cell material from the loop by dipping the loop in an alcohol sand bottle. Then the loop is heated red hot in the Bunsen burner.

7. Permit the slide to dry. Do not heat it in any way to hasten the process, since the cells will

Figure 5.4 *(a)* Staining, *(b)* washing, and *(c)* blotting a simple stain. From John P. Harley and Lansing M. Prescott, *Laboratory Exercises in Microbiology*, 5th ed. Copyright © 2002 The McGraw-Hill Companies. All Rights Reserved. Reprinted by permission.

Simple Staining Procedure

(a)

(b)

(c)

Staining loop

Sink or suitable receptacle

Stain

Staining bottle

Water

Wash bottle

Gentle blotting

15. **Important:** Wipe off the oil from the immersion lens with lens paper before storing the microscope.

Negative Stain

The negative stain can be used to observe capsules or inclusion bodies. However, in this exercise the negative stain will be used to compare the appearance of the same organisms using the two staining procedures.

Materials

Same cultures used for simple stain
Bottle of India ink

Procedure

1. Place a drop of water on a clean slide and add organisms with a loop until the drop is cloudy.
2. Sterilize the loop, mix a loopful of India ink into the drop, and spread the mixture out into a thin film.
3. Sterilize the loop.
4. Let dry and examine under the microscope. Bacteria can be seen as clear areas on a black background.
5. Record your results.

EXERCISE 5

Laboratory Report: Simple Stains:
Positive and Negative Stains

Results

1. Simple stain

	Staphylococcus	*Bacillus*	*Micrococcus*	*Streptococcus*
Draw shape and arrangement				

2. Negative stain

Questions

1. What are the advantages of a simple stain over a wet mount?

2. Do you need more or less light when viewing a stained preparation compared to a wet mount?

3. What can you observe in a wet mount that cannot be seen in a stained preparation?

4. How does the appearance of the negative stain compare to the appearance of the simple stain?

5. How many μm are in a millimeter (mm)? _____
   How many μm are in a meter (m)? _____

EXERCISE 6

Differential and Special Stains

Differential Stains: Gram Stain and Acid-Fast Stain

Getting Started

Differential stains usually involve at least two dyes and are used to distinguish one group of organisms from another. For example, the Gram stain determines whether organisms are Gram positive or Gram negative.

Gram Stain

The Gram stain is especially useful as one of the first procedures employed in identifying organisms. It reveals not only the morphology and the arrangement of the cells, but also information about the cell wall.

In the late 1800s, Christian Gram devised the staining procedure when trying to stain bacteria so that they contrasted with the tissue sections he was observing. Many years later, it was found that purple (Gram-positive) bacteria had thick cell walls of **peptidoglycan,** while pink (Gram-negative) bacteria had much thinner cell walls of peptidoglycan surrounded by an additional lipid membrane. The thick cell wall retains the purple dye in the procedure, but the thin wall does not (table 6.1).

In the Gram stain, a bacterial smear is dried and then heat-fixed to cause it to adhere to the glass slide (as in the simple stain). It is then stained with crystal violet dye, the primary stain, which is rinsed off and replaced with an iodine solution. The iodine acts as a **mordant**—that is, it binds the dye to the cell. The smear is then decolorized with alcohol and **counterstained** with safranin. In Gram-positive organisms, the purple crystal-violet dye, complexed with the iodine solution, is not removed by the alcohol, and thus the organisms remain purple. On the other hand, the purple stain is removed from Gram-negative organisms by the alcohol, and the colorless cells take up the red color of the safranin counterstain.

Note: Many clinical laboratories use a 50/50 mixture of alcohol and acetone because it destains faster than 95% alcohol. If the instructor would rather not use acetone, 95% alcohol is just as effective, but the stain must be decolorized longer (up to 30 seconds).

Special Notes to Improve Your Gram Stains

1. Gram-positive organisms can lose their ability to retain the crystal violet complex when they are old. This can happen when a culture has only been incubating 18 hours—the genus *Bacillus* is especially apt to become Gram negative. Use young, overnight cultures whenever possible. It is interesting to note that Gram-positive organisms can appear Gram negative, but Gram-negative organisms almost never appear Gram positive.
2. Another way Gram-positive organisms may appear falsely Gram negative is by over decolorizing in the Gram-stain procedure. If excessive amounts of acetone/alcohol are used, almost any Gram-positive organism will lose the crystal violet stain and appear Gram negative.
3. If you are staining a very thick smear, it may be difficult for the dyes to penetrate properly. This is not a problem with broth cultures, which are naturally quite dilute, but be careful not to make the suspension from a colony in a drop of water too thick.
4. When possible, avoid making smears from inhibitory media such as eosin methylene blue (EMB) because the bacteria frequently give variable staining results and can show atypical morphology.

Table 6.1 Appearance of the Cells After Each Procedure

	Gram +	Gram −
Crystal violet	Purple	Purple
Iodine	Purple	Purple
Alcohol	Purple	Colorless
Safranin	Purple	Pink

5. The use of safranin in the Gram stain is not essential. It is simply used as a way of dying the colorless cells so they contrast with the purple. For those who are color-blind and have difficulty distinguishing pink from purple, other counterstain dyes can be tried.

Definitions

Counterstain. A stain used to dye unstained cells a contrasting color in a differential stain.

Mordant. A substance that increases the adherence of a dye.

Peptidoglycan. The macromolecule making up the cell wall of most bacteria.

Vegetative cell. A cell that has not formed spores or other resting stages.

Objectives

1. To learn the Gram-stain procedure.
2. To learn to distinguish Gram-positive organisms from Gram-negative organisms.

References

Gerhardt, Philip, ed. *Manual for general and molecular bacteriology.* Washington, D.C.: American Society for Microbiology, 1993, paperback.

Murray, Patrick et al. *Manual of clinical microbiology.* 8th ed. Washington, D.C.: ASM Press, 2003.

Nester et al. *Microbiology: A human perspective,* 6th ed. The McGraw-Hill Companies, 2009. Chapter 3, Section 3.2.

Materials

Staining bottles of the following:
  crystal violet
  iodine
  acetone/alcohol or 95% alcohol
  safranin
Clothespin or forceps
Staining bars
Disposable gloves

Overnight cultures growing on TS agar slants of the following:
  Escherichia coli
  Bacillus subtilis
  Staphylococcus epidermidis
  Streptococcus mutans
  Micrococcus luteus

Procedure for Gram Stain

1. Put two drops of water on a clean slide. In the first drop, make a suspension of the unknown organism to be stained just as you did for a simple stain (see preparing a smear in exercise 5). In the second drop, mix together a known Gram-positive organism and a known Gram-negative organism. This mixture is a control to ensure that your Gram-stain procedure (figure 6.1) gives the proper results. Heat-fix the slide.

2. Place the slide on a staining bar across a sink (or can). Alternatively, hold the slide with a clothespin or forceps over a sink.

3. Flood the slide with crystal violet until the slide is completely covered. Leave it on for 6–30 seconds and then discard into the sink. The timing is not critical. Rinse the slide with water from a wash bottle or with gently running tap water.

4. Flood the slide with iodine for about 12–60 seconds and then wash with water.

5. Hold the slide at a 45° angle and carefully drip acetone/ethanol, the decolorizer, over it until no more purple dye runs off. Immediately wash the slide with tap water. Thicker smears may take longer than thinner ones, but acetone/alcohol should usually be added for 1–2 seconds and no more than 5 seconds. Timing is critical in this step.

6. Flood the slide with safranin and leave it on for 10–30 seconds—timing is not important. Wash with tap water. Safranin is a counterstain because it stains the cells that have lost the purple dye.

7. Blot the slide carefully with a paper towel to remove the water, but do not rub from side to side. When it is completely dry, observe the slide under the microscope. Remember that

Figure 6.1 Gram-stain procedure.

(a) Apply Crystal violet to smear; 6–30 seconds

(b) Rinse with water for 5 seconds.

(c) Cover with Gram's iodine; 12–60 seconds.

(d) Rinse with water for 5 seconds.

(e) Decolorize 1–5 seconds.

(f) Rinse with water for 5 seconds.

(g) Counterstain with safranin; 10–30 seconds.

(h) Rinse with water for 5 seconds.

(i) Blot dry with a paper towel.

you must use the oil immersion lens to observe bacteria. Compare your stain to the control mixture on the same slide and with color plate 7.

8. Describe the appearance of your stained bacteria in the Results section of the Laboratory Report.

9. Be sure to remove the immersion oil from the lens with lens paper before storing the microscope.

Acid-Fast Stain (optional)

Getting Started

The acid-fast stain is useful for identifying bacteria with a waxy lipid cell wall. Most of these organisms are members of a group of bacteria called mycobacteria. Although there are many harmless bacteria in this group, it also includes *Mycobacterium tuberculosis*, which is the cause of tuberculosis in humans. These organisms have a Gram-positive cell wall structure, but the lipid in the cell wall prevents staining with the Gram-stain dyes.

In the Ziehl-Neelsen (Kinyoun modification) acid-fast stain procedure, the dye carbolfuchsin stains the waxy cell wall. Once the lipid-covered cell has been dyed, it cannot be decolorized easily—even with alcohol containing HCL (called acid-alcohol). Nonmycobacteria are also dyed with the carbolfuchsin but are decolorized by acid-alcohol. These colorless organisms are stained with methylene blue, so they contrast with the pink acid-fast bacteria that were not decolorized.

The reason this stain is important is that one of the initial ways tuberculosis is diagnosed is by the presence of *Mycobacterium* in a patient's sputum. (Sputum is a substance that is coughed up from the lungs and contains puslike material.) Tuberculosis is a very serious disease worldwide and is now seen in the United States after decreasing for about 80 years. Since the process of finding acid-fast organisms in sputum is quite difficult and time-consuming, this test is usually performed in state health laboratories.

Objectives

1. To become familiar with acid-fast organisms.
2. To prepare an acid-fast stain.

References

Gerhardt, Philip, ed. *Manual for general and molecular bacteriology.* Washington, D.C.: American Society for Microbiology, 1993, paperback.

Nester et al. Microbiology: A human perspective, 6th ed. The McGraw-Hill Companies, 2009. Chapter 3, Section 3.2.

Materials

Culture of the following:

Mycobacterium smegmatis mixed with *Staphylococcus epidermidis*

Carbolfuchsin in staining bottles

Methylene blue in staining bottles

Acid-alcohol in staining bottles

Beaker

Metal or glass staining bars

Procedure for Acid-Fast Stain (Kinyoun Modification)

1. Prepare a smear of the material and heat-fix (see exercise 5).
2. Cover the smear with carbolfuchsin and stain for 3–5 minutes. Do not heat (figure 6.2).
3. Rinse with water.
4. Decolorize with acid-alcohol for 10–30 seconds.
5. Rinse with water.
6. Counterstain with methylene blue for 20–30 seconds.
7. Rinse with water.
8. Blot dry carefully and examine under the oil immersion lens.
9. Record results.

Figure 6.2 Acid-fast stain procedure.

(a) Apply carbolfuchsin to smear for 5 minutes.

(b) Rinse with water.

(c) Decolorize with acid-alcohol; 10–30 seconds.

(d) Rinse with water.

(e) Counterstain with methylene blue; 20–30 seconds.

(f) Rinse with water.

(g) Blot dry with a paper towel.

Special Stains: Capsule, Endospore, Storage Granules, and Flagella

Getting Started

Although bacteria have few cell structures observable by light microscopy when compared to other organisms, some have capsules, endospores, storage granules, or flagella. The following procedures will enable you to see these structures.

Objectives

1. To become familiar with various structures and storage products of bacteria.
2. To learn various staining procedures for these structures.

References

Gerhardt, Philip, ed. *Manual for general and molecular bacteriology.* Washington, D.C.: American Society for Microbiology, 1993, paperback.

Murray, Patrick et al. *Manual of clinical microbiology.* 8th ed. Washington, D.C.: ASM Press, 2003.

Nester et al. *Microbiology: A human perspective,* 6th ed. The McGraw-Hill Companies, 2009. Chapter 3, Section 3.2.

Capsule Stain (optional)

A capsule is a somewhat gelatinous coating surrounding the cell. It can consist of amino acids or carbohydrates and it can protect the bacterium from engulfment by white blood cells. The ability to produce a capsule frequently depends on the availability of certain sugars. *Streptococcus mutans*, for example, produces a capsule when growing on sucrose but not when growing on glucose.

Materials

Culture of the following:
  Klebsiella or other organism with a capsule growing on a slant
India ink

Procedure for Capsule Stain

1. Make a suspension of the organism in a drop of water on a clean slide.
2. Put a drop of India ink next to it.
3. Carefully lower a coverslip over the two drops so that they mix together. There should be a gradient in the concentration of the ink.
4. Examine the slide under the microscope and find a field where you can see the cells surrounded by a halo in a black background.
5. Drop slides in a beaker or can of boiling water and boil for a few minutes before cleaning. This is necessary because the bacteria are not killed in the staining process.
6. Record results.

Endospore Stain (optional)

Some organisms such as *Bacillus* and *Clostridium* can form a resting stage called an endospore, which protects them from heat, chemicals, and starvation. When the cell determines that conditions are becoming unfavorable due to a lack of nutrients or moisture, it forms an endospore. When conditions become favorable again the spore can germinate and the cell can continue to divide. The endospore is resistant to most stains so special staining procedures are needed.

Materials

Culture of the following:
  Bacillus cereus on a nutrient agar slant after 3–4 days incubation at 30°C
Malachite green in staining bottles
Safranin in staining bottles
Metal or glass staining bars
Beaker or can

Procedure for Endospore Stain

1. Prepare a smear on a clean slide and heat-fix.
2. Add about an inch of water to a beaker and bring it to a boil.

Figure 6.3 Procedure for endospore stain.

(a) Apply malachite green to saturate paper, and steam for 5 minutes.

(b) Remove paper, cool, and rinse with water for 30 seconds.

(c) Counterstain with safranin for 30 seconds.

(d) Rinse with water for 30 seconds.

(e) Blot dry with a paper towel.

Figure 6.4 Appearance of endospores stained with spore stain and Gram stain. *Note:* The *Bacillus* bacteria frequently lose their ability to stain Gram positive.

Spore Stain of *Bacillus* with Malachite Green

Vegetative cells (pink)   Sporulated cells

Endospores (green)

Gram Stain of *Bacillus*

Vegetative cells (purple)   Sporulated cells

Gram + rods   Endospores not stained

3. Place two short staining bars over the beaker and place the slide on them.
4. Tear a piece of paper towel a little smaller than the slide and lay on top of the smear. The paper prevents the dye from running off the slide.
5. Flood the slide with malachite green and steam for 5 minutes. Continue to add stain to prevent the dye from drying on the slide (figure 6.3).
6. Decolorize with water for about 30 seconds by flooding with water or holding under gently running tap water. The **vegetative cells** (dividing cells) lose the dye, but the endospores retain the dye.
7. Counterstain with safranin for about 30 seconds and then wash with tap water for 30 seconds. Blot dry carefully.
8. Observe with the oil immersion lens. The endospores appear green and the vegetative cells appear pink. Sometimes the endospore is still seen within the cell, and its shape and appearance can be helpful in identifying the organism. In other cultures, the endospores may be free because the cells around them have disintegrated (figure 6.4).
9. Record results.
10. Prepare and observe a Gram stain of the same culture (optional).

Note: When bacteria containing endospores are Gram stained, the endospores do not stain and the cells appear to have holes in them (figure 6.4).

Storage Granules Stain (optional)

Many organisms can accumulate material as storage granules that are abundant in their environment for use in the future. For example, phosphate can be stored as metachromatic granules (also called volutin granules). When organisms containing these granules are stained with methylene blue, the phosphate granules are stained a darker reddish blue.

Materials

Culture of the following:

    Spirillum grown in nutrient broth

Methylene blue in staining bottles

Procedure for Storage Granules Stain

1. Prepare a smear from the broth. It might be helpful to remove the organisms from the bottom of the tube with a capillary pipet. Place a drop on a clean slide. Dry and heat-fix.
2. Flood the slide with methylene blue for about 20–30 seconds.
3. Wash with tap water and blot dry.
4. Observe with the oil immersion lens. The metachromatic granules should appear as dark reddish-blue bodies within the cells.
5. Record results.

Flagella Stain (optional)

Some bacteria have flagella (flagellum, singular) for motility. Their width is below the resolving power of the microscope so they cannot be seen in a light microscope (the flagella seen at each end of *Spirillum* in a wet mount is actually a tuft of flagella). Flagella can be visualized if they are dyed with a special stain that precipitates on them, making them appear much thicker. The arrangement of the flagella on bacteria is usually characteristic of the organism and can aid in identification. There are three basic kinds of arrangement of flagella: **peritrichous, polar,** and **polar tuft.**

Definitions

Peritrichous.   The surface of the organism is covered with flagella, such as *E. coli.*

Polar.   A single flagellum at one or both ends of the organism, such as *Pseudomonas.*

Polar tuft (lophotrichous).   A tuft of flagella at one or each end of an organism.

Materials

Stained demonstration slides of *Escherichia coli* and *Pseudomonas*

Procedure for Flagella Stain

1. Observe slides with stained flagella of several organisms and note the pattern of flagella. It is difficult to perform this staining procedure, so prestained slides are recommended.
2. Record results.

NOTES:

EXERCISE

6

Laboratory Report: Differential and Special Stains

Results

	Gram Reaction	Arrangement (sketch)
E. coli		
B. subtilis		
S. epidermidis		
Streptococcus mutans		
M. luteus		

Optional Stains	Organism	Appearance
Acid-fast		
Capsule		
Endospore		
Storage granules		
Flagella		

Questions

1. What is the function of each one of the Gram-stain reagents?
   a.

   b.

   c.

   d.

2. Give two reasons Gram-positive organisms sometimes appear Gram negative.

3. What is the purpose of using a control in the Gram stain?

4. What is a capsule?

5. What are storage granules, and why are they important to the cell?

6. How does an endospore appear (draw and indicate color):

   a. when Gram stained?

   b. when spore stained?

7. What is another way you can determine whether or not an organism is motile besides observing a flagella stain?

8. Why can't you Gram stain an acid-fast organism?

NOTES:

*I*NTRODUCTION to Microbial Growth

The ability to grow bacteria is very important for studying and identifying them. Organisms in the laboratory are frequently grown either in a broth culture or on a solid agar medium. A broth culture is useful for growing large numbers of organisms. Agar medium is used in a petri dish when a large surface area is important, as in a streak plate. On the other hand, agar medium in tubes (called slants) is useful for storage because the small surface area is not as easily contaminated and the tubes do not dry out as fast as plates. You will be able to practice using media in all these forms (figure I.7.1).

You will also use different kinds of media in this section. Most media are formulated so that they will support the maximum growth of various organisms, but other media have been designed to permit the growth of desired organisms and inhibit others (selective). Still other media have been formulated to change color or in some other way distinguish one bacterial colony from another (differential). These media can be very useful when trying to identify an organism.

It is also important to know how to count bacteria. You will have an opportunity to learn about several techniques and their advantages and disadvantages.

In these exercises no pathogenic organisms are used, but it is very important to treat these cultures as if they were harmful because you will be prepared to work safely with actual pathogens. Also, almost any organism can cause disease if there are large numbers in the wrong place.

Figure I.7.1 Various media in different forms.

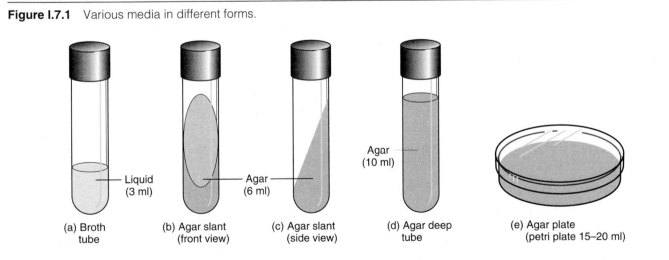

(a) Broth tube — Liquid (3 ml)

(b) Agar slant (front view) — Agar (6 ml)

(c) Agar slant (side view)

(d) Agar deep tube — Agar (10 ml)

(e) Agar plate (petri plate 15–20 ml)

NOTES:

EXERCISE 7

Chemically Defined, Complex, Selective, and Differential Media

Getting Started

Microbiologists have developed several different types of culture media for a variety of purposes. If a medium is totally made up of specific amounts of chemicals it is called a **chemically defined medium.** If, however, it contains such mixtures as yeast extract or enzymatic digests of protein, it is termed a **complex medium (or undefined medium)** because the exact amount and kinds of large organic molecules are not known.

Complex media are frequently called **rich** (or **enriched) media.** These media tend to support the growth of bacteria better because they contain more preformed nutrients and the organisms do not have to use energy and materials to synthesize the compounds supplied in the medium. Many bacteria can grow only on this kind of medium because they cannot synthesize all the necessary components needed for growth and must be provided with preformed amino acids, vitamins, and other growth factors. Some organisms require only one or two vitamins or amino acids, while other bacteria need many complex growth factors; these are termed *fastidious.*

Two other valuable types of media are **selective media** and **differential media.** Frequently, it is important to isolate one organism in a mixture of bacteria. Normal biota can contaminate the culture; for example, a wound culture may be mixed with large amounts of *Staphylococcus* normally found on the skin. It can be difficult to isolate or even detect the pathogenic organism among all the nonpathogenic organisms present. Selective media have been designed to permit some bacteria to grow but not others, so that certain bacteria can be isolated even if they constitute only a small percentage of the population.

Differential media are also useful for isolating and identifying bacteria. By observing the appearance of colonies growing on this agar, it is possible to determine such characteristics as whether or not they can ferment certain sugars.

In this exercise, you will have an opportunity to observe the growth of organisms on three different media.

Glucose Salts Agar This is a simple chemically defined medium. Only organisms that can make all their cellular components from glucose and inorganic salts are able to grow on it.

Trypticase Soy Agar This is a rich, complex medium made from an enzymatic digest of protein and soy product. Organisms that require vitamins or other growth factors are able to grow on it.

Eosin Methylene Blue (EMB) Agar This is a selective medium permitting the growth of Gram-negative enteric rods and inhibiting the growth of Gram-positive bacteria. In addition, the medium is also differential because it contains the sugar lactose. Organisms that can ferment lactose produce purple colonies, and those that cannot produce white or very light pink colonies. The colonies of *Escherichia coli*, a lactose fermenter, are dark purple. They also give the medium a distinctive metallic green sheen caused by the large amounts of acid produced. The colonies of *Enterobacter*, also a lactose fermenter, usually are more mucoid with pink to purple centers. (Mucoid colonies have a slimy appearance.)

Definitions

Complex medium. Also termed an *undefined medium*. A medium in which the exact amounts of chemical components and their composition are unknown because it is made of extracts or enzymatic digests of meat, plants, or yeast.

Chemically defined medium. A synthetic medium composed of inorganic salts and usually a carbon source such as glucose.

Differential medium. Medium permitting certain organisms to be distinguished from others by the appearance of their colonies.

Rich (or enriched) medium. A medium, usually complex, containing many growth factors. Sometimes blood or additional vitamins are added to permit the growth of organisms with special requirements.

Selective medium. Medium formulated to permit the growth of certain bacteria but not others.

Undefined medium. Also termed a *complex medium*. A medium in which the exact amounts of chemical components and their composition are unknown because it is made of extracts or enzymatic digests of meat, plants, or yeast.

Objectives

1. To compare the growth of organisms on a chemically defined medium and a rich medium.
2. To compare the growth of organisms on a selective medium and a differential medium.
3. To understand the relationship between the growth of an organism and the composition of the medium.

Reference

Nester et al. *Microbiology: A human perspective,* 6th ed. The McGraw-Hill Companies, 2009. Chapter 4, Section 4.1.

Materials

Cultures growing in trypticase soy broth of the following:

   Escherichia coli

   Staphylococcus epidermidis

   Pseudomonas aeruginosa

   Enterobacter aerogenes

Media per team

   Trypticase soy (TS) agar plate

   Glucose salts agar plate

   EMB agar plate

Procedure

First Session

1. With a marking pen, divide the bottom of the petri plates into quadrants. Label the plates with your name and date. Label each quadrant with the organism as shown in figure 7.1.
2. Inoculate each quadrant of the plate with a loopful of the culture using a wavy line.
3. Invert and incubate at 37°C for 48 hours.

Second Session

1. Observe and compare the growth on the three plates.
2. Record the results.

Figure 7.1 A diagram of the labeled media plates.

Trypticase soy (TS)   Glucose mineral   Eosin methylene blue (EMB)

EXERCISE

7

Laboratory Report: Chemically Defined, Complex, Selective, and Differential Media

Results

	Glucose Salts	TS	EMB
			lac–/lac+
Escherichia coli			
Staphylococcus			
Pseudomonas			
Enterobacter			

Indicate the amount of growth:
    0 = no growth
    + = slight growth
    ++ = good growth
  Lac – = no lactose fermentation on EMB
  Lac + = lactose fermentation on EMB

Questions

1. Which organisms could not grow on the glucose salts medium? Which organisms could grow on it?

2. Which organisms do not require any growth factors?

3. Of the organisms that could grow on both TS agar and glucose salts agar, did some organisms grow better on the TS agar than on the glucose salts? Can you propose a reason?

4. Which organisms could grow on the EMB agar?

5. Which organisms could ferment lactose?

6. Could you differentiate *E. coli* from other organisms growing on the EMB agar? How?

7. In general, EMB agar selects for what kind of organisms?

8. What kinds of organisms does EMB agar differentiate?

EXERCISE 8

Quantification of Microorganisms

Getting Started

It is frequently important to count bacteria. For example, you may want to know the number of bacteria in a sample of raw chicken or the number of bacteria per ml of water in a swimming pool. Special techniques have been devised to enumerate bacteria, each with advantages and disadvantages. Three common methods are discussed: plate count, direct count, and the turbidometric method.

Plate Count   This method is based on the premise that each **viable** bacterium produces a colony when growing on an agar plate. A sample of the material to be counted is suspended in liquid and placed in an empty petri plate. Melted, cooled agar is then added to the plate and mixed with the inoculum. After incubation, each organism produces a colony in the agar that can then be counted. The plate count is used very frequently, but it has advantages and disadvantages that should be considered prior to use. Some of these are discussed next.

1. Bacteria are usually present in very large numbers—an overnight broth culture of **E. coli** can easily contain one billion cells/ml However, the maximum number of colonies that can be accurately counted on a plate is usually set at 300. Therefore, most samples must be diluted to low enough numbers that the plates will have distinct colonies that can be counted. Since it is usually not possible to know exactly how diluted to make a sample to obtain a countable plate, several different dilutions must be plated.

2. Some bacteria tend to stick together; therefore, sometimes two or more bacteria give rise to one colony. This gives results of a lower number than are actually present.

3. If a sample has many different kinds of bacteria, it is not possible to have a medium or conditions that support all their various necessary growth conditions. Soil, for example, may contain organisms that will not grow unless the temperature is above 50°C; in contrast, other organisms are inhibited at these temperatures. These problems must be considered when a sample of mixed bacteria is enumerated.

4. It can take at least 24 hours to obtain the results of a plate count.

5. The plate count method does have two advantages over other methods, however. Only viable organisms are counted, which are the ones usually considered important. Also, samples with small numbers—considered insufficient for other methods—can be counted.

Direct Count   In this method of counting organisms, a suspension of bacteria is placed on a slide that has been ruled into squares and is designed to hold a specific volume of liquid. By counting the bacteria that appear on the grid areas, the number of organisms in the sample can be calculated. It is a much faster test than the plate count, but it does have several drawbacks. First, there must be about 1×10^7 organisms/ml before there are enough to be seen, and second, viable and nonviable organisms appear the same under a microscope.

Turbidometric Method   In this method of counting, a spectrophotometer is used to measure the **turbidity** or **optical density (O.D.)** of bacteria in a broth. The more bacteria, the cloudier the broth and the higher the O.D. In this method, you must first correlate plate counts with O.D. readings. This must be done with each strain of bacteria because organisms are different sizes. For instance, an O.D. reading of 0.2 for a broth culture of one *E. coli* strain is equal to 1×10^8 cells/ml. The same number of another organism would have a different O.D. Once the correlation between O.D. and plate counts has been determined, it can be used as an extremely convenient method of determining numbers of organisms. This method is used to determine the generation time in exercise 10. Other methods of enumerating bacteria are discussed in exercise 32, which is on water analysis.

The plate count method is used in this exercise to count the number of organisms in two broth cultures: one turbid (sample A) and the other with no visible turbidity (sample B). There are two methods of preparing plate counts: pour plates and spread plates. In the pour plate method (as just described), a sample is mixed with melted agar in a petri plate and colonies appearing in and on the agar are counted. In the spread plate method, a small sample is placed on the surface of the agar plate and spread with a bent glass rod so that all the colonies appear on the surface of the plate. The spread plate option for this exercise is found in Appendix 4.

Definitions

Optical density (O.D.). A measure of the amount of turbidity. Frequently called *absorbance*.

Serial dilution. Preparing a dilution in steps instead of one dilution.

Turbidity. Cloudiness.

Viable (bacteria). Capable of growing and dividing.

Objectives

1. To enumerate bacteria using the plate count method.
2. To understand the use of dilutions.
3. To gain insight into the number of organisms that can be present in a clear liquid and in a turbid liquid.

Reference

Nester et al. *Microbiology: A human perspective,* 6th ed. The McGraw-Hill Companies, 2009. Chapter 4, Section 4.8.

Note: When serially transferring concentrated suspensions to less concentrated suspensions, a new pipet must be used for each transfer to prevent carryover of organisms. (In this exercise, the instructor may choose to use each pipet twice to conserve materials.) However, when going from less concentrated to more concentrated suspensions, the same pipet may be used with no significant error because only a few organisms will be added to the much larger sample.

Materials

For teams using sample A (turbid suspension)

99-ml water blank, 1

9.9-ml water blanks, 2

9.0-ml water blanks, 3

1-ml pipets, 7

TS agar deeps, 4

Sterile petri dishes, 4

Suspension A: An overnight TS broth culture (without shaking) of *E. coli* diluted 1:1 with TS broth

---

Procedure for Team A

First Session

1. Observe whether the suspension is cloudy or clear.
2. Label all water blanks with the dilution, as shown in figure 8.1.
3. Melt 4 TS agar deeps and hold at 50°C. It is very important not to let the deeps cool much lower than 50°C because the agar will harden and will have to be heated to boiling (100°C) before it will melt again (figure 8.2).
4. Make **serial dilutions** of the bacterial suspension.
   a. Mix the bacterial suspension by rotating between the hands and transfer 1.0 ml of the suspension to the 99-ml water blank labeled 10^{-2}. Discard pipet.
   b. Mix well and transfer 0.1 ml of the 10^{-2} dilution to the 9.9-ml water blank labeled 10^{-4}. Discard pipet.
   c. Mix well and transfer 0.1 ml of the 10^{-4} dilution to the 9.9-ml tube labeled 10^{-6}. Discard pipet.
   d. Mix and transfer 1.0 ml of the 10^{-6} dilution to the 9.0-ml tube labeled 10^{-7}. (Note change from 0.1 ml to 1.0 ml.) Discard pipet.

Figure 8.1   Dilution scheme for Team A.

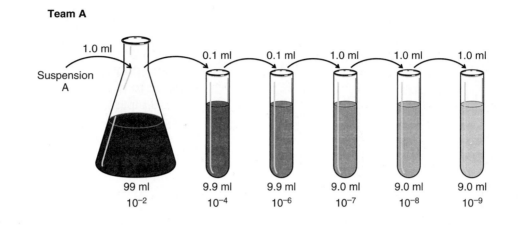

Team A

1.0 ml — Suspension A

| 99 ml | 9.9 ml | 9.9 ml | 9.0 ml | 9.0 ml | 9.0 ml |
| 10^{-2} | 10^{-4} | 10^{-6} | 10^{-7} | 10^{-8} | 10^{-9} |

0.1 ml   0.1 ml   1.0 ml   1.0 ml   1.0 ml

Figure 8.2   Melting and pouring agar deeps.

(a) Boiling water bath

(b) 50°C water bath — Thermometer, Test tube rack

(c) Wipe tube with paper towel.

(d) Flame the top of the tube after removing cap.

(e) Pour the agar into the petri dish bottom.

(f) After cooling, store in an inverted position. — Agar

e. Mix and transfer 1.0 ml of the 10^{-7} dilution to the 9.0-ml tube labeled 10^{-8}. Discard pipet.

f. Mix and transfer 1.0 ml of the 10^{-8} dilution to the 9.0-ml tube labeled 10^{-9}. Discard pipet.

5. Place samples of the dilutions 10^{-9}, 10^{-8}, 10^{-7}, and 10^{-6} into sterile, labeled petri plates as follows:

a. Mix the 10^{-9} dilution, and transfer 1.0 ml into a sterile petri plate labeled 10^{-9}. Add a tube of melted agar (wipe off the outside of the tube before pouring) (figure 8.2) and swirl gently by moving the plate in a figure eight pattern on the bench. Do not discard pipet.

b. Mix the 10^{-8} dilution, and with the same pipet, transfer 1.0 ml to the petri plate labeled 10^{-8}. Add melted agar and mix.

c. Mix the 10^{-7} dilution again, and using the same pipet, transfer 1.0 ml to the petri plate labeled 10^{-7}. Add melted agar and mix.

d. Mix the 10^{-6} dilution, and using the same pipet, transfer 1.0 ml to the petri plate labeled 10^{-6}. Add melted agar and mix. Discard pipet.

6. Invert the plates after you are sure the agar has hardened (about 5 minutes), and incubate at 37°C.

Materials

For teams using sample B (a nonturbid suspension)

99-ml water blank, 1

9.9-ml water blank, 1

9.0-ml water blanks, 4

1-ml pipets, 5

TS agar deeps, 4

Sterile petri dishes, 4

Suspension B: A nonturbid suspension of bacteria

Procedure for Team B

First Session

1. Observe whether the suspension is cloudy or clear.
2. Label all water blanks and plates with the dilution (figure 8.3).
3. Melt 4 TS agar deeps and hold at 50°C. It is very important not to let the agar deeps cool much lower than 50°C because the agar will harden and will have to be heated to boiling (100°C) before it will melt again.
4. Make serial dilutions of the bacterial suspension.
   a. Mix the bacterial suspension and transfer 1.0 ml to the 99-ml water blank labeled 10^{-2}. Discard pipet.
   b. Mix well and transfer 0.1 ml of the 10^{-2} dilution to the 9.9-ml water blank labeled 10^{-4}. (Note that this is

the only time you use a 0.1 ml-sample.) Discard pipet.
   c. Mix well and transfer 1.0 ml of the 10^{-4} dilution to the 9.0-ml tube labeled 10^{-5}. Discard pipet.
   d. Mix and transfer 1.0 ml of the 10^{-5} dilution to the 9.0-ml tube labeled 10^{-6}. Discard pipet.
   e. Mix and transfer 1.0 ml of the 10^{-6} dilution to the 9.0-ml tube labeled 10^{-7}. Discard pipet.
5. Place samples of the dilutions 10^{-7}, 10^{-6}, 10^{-5}, and 10^{-4} into the pour plates as follows.
   a. Mix the 10^{-7} dilution tube and remove 1 ml to the petri plate labeled 10^{-7}. Pour the melted, cooled agar into the plate (wipe off any water on the outside of the tube before pouring the agar) (figure 8.2). Mix by gently moving the plate in a figure eight on the bench. Do not discard the pipet.
   b. Repeat the procedure for the 10^{-6}, 10^{-5}, and 10^{-4} dilutions using the same pipet.
6. After the agar has hardened (about 5 minutes) invert the plates and incubate at 37°C.

Second Session for Both A and B Teams

1. Count the colonies in the agar plates. Use a marking pen on the bottom of the plate to dot the colonies as you count them. Colonies growing in the agar tend to be lens-shaped and smaller than those growing on the surface, although all are counted equally. If there are more than 300 colonies on the plate, label it TNTC—too numerous to count.

Figure 8.3 Dilution scheme for Team B.

Team B

Suspension B

99 ml — 10^{-2} 9.9 ml — 10^{-4} 9.0 ml — 10^{-5} 9.0 ml — 10^{-6} 9.0 ml — 10^{-7}

1.0 ml   0.1 ml   1.0 ml   1.0 ml   1.0 ml

2. Choose the plate that has between 30 and 300 colonies (less than 30 gives results with a high sample error). Calculate the number of organisms/ml using the following formula:

number of organisms on the plate × 1/sample volume × 1/dilution = the number of organisms/ml in the original suspension

Note: In this exercise the volume of all the samples is 1.0 ml.

3. Record your results. Post your results on the blackboard so that average numbers of organisms/ml for Suspension A and Suspension B can be calculated.

Understanding Dilutions

(See also Appendix 2.)

1. To make a dilution use the following formula:

sample/(diluent + sample) = the dilution

Example 1: How much is a sample diluted if 1 ml is added to 9.0 ml of water (the water is sometimes called a diluent)?

1/(1 + 9) = 1/10 (also expressed 10^{-1})

Example 2: How much is a sample diluted if 0.1 ml is added to 9.9 ml of water?

$$0.1/(0.1 + 9.9) = 0.1/10$$
$$= 1:100 \text{ or } 10^{-2}$$

2. When a sample is serially diluted, multiply each dilution together for the final dilution. The final dilution in tube B is 1:100 or 10^{-2}.

3. To calculate the number of organisms in the original sample use this formula:

The number of organisms/ml in the original sample = number of colonies on plate × 1/volume of sample × 1/dilution

1 ml   1 ml

9 ml   9 ml
Tube A   Tube B
10^{-1} 10^{-2}
(1/10)   (1/100)

Example 3: Suppose you counted 120 organisms on a plate diluted 10^{-2}. The sample size was 0.1 ml.

Solution: 120 (number of organisms on plate) × 1/0.1 × $1/10^{-2}$ = 120 × 10 × 100
= 120 × 10^3 or 1.2 × 10^5 organisms/ml

Example 4: Suppose you counted 73 colonies on the plate marked 10^{-6}. If the sample size was 1.0 ml then

$73 × 1/1.0 × 1/10^{-6} = 73 × 1 × 10^6 = 73 × 10^6$ or 7.3 × 10^7 organisms/ml in the original suspension

(It is important to label the answer "per ml.")

NOTES:

EXERCISE

8

Laboratory Report: Quantification of Microorganisms

Results

1. Which plate (dilution) had between 30 and 300 colonies?

2. How many colonies did you count?

3. How many organisms/ml were in the original suspension that you counted (from questions 1 and 2)? Show calculations. Record on blackboard.

	Suspension A	Suspension B
Appearance of broth		
Number of organisms/ml (class average)		
Number of organisms/ml (your data)		

Questions

1. From these results, about how many organisms/ml can be in a cloudy broth? (Show calculations.)

2. From these results, about how many organisms/ml can be in a clear broth without showing any sign of turbidity? (Show calculations.)

3. What are two sources of error in this procedure?

4. If you serially dilute a sample with three 1:10 dilutions, what is the dilution of the last tube?

5. If you add 1.0 ml to 99 ml of water, what is the dilution of the sample?

6. If you had a solution containing 6,000 organisms/ml, how could you dilute and plate a sample so that you had a countable plate?

*I*NTRODUCTION to the Environment and Microbial Growth

An organism cannot grow and divide unless it is in a favorable environment. Environmental factors include temperature, availability of nutrients, moisture, oxygen, salinity, osmotic pressure, and presence of toxic products.

Each bacterial species has its own particular set of optimal conditions that allows maximum growth. These conditions probably reflect the environment in which the organism grows and competes with other organisms. For example, organisms found in a hot spring in Yellowstone National Park require a much higher temperature for growth in the laboratory than organisms isolated from the throat. In the next two exercises, we examine the effects of temperature and atmosphere (oxygen) on the growth of bacteria. (Osmotic pressure is examined in exercise 13.)

NOTES:

EXERCISE

9

Aerobic and Anaerobic Growth

Getting Started

All the animals we are familiar with, including humans, have an absolute requirement for oxygen. It seems rather surprising then that there are groups of organisms that cannot grow or are even killed in the presence of oxygen. Still other kinds of organisms can grow either with or without oxygen. These three groups are classified as follows:

1. *Obligate aerobes* Organisms that have an absolute requirement for oxygen. *Micrococcus,* a member of the skin flora, and *Pseudomonas,* a soil organism (and occasional pathogen), are obligate aerobes.

2. *Obligate anaerobes* Organisms that cannot grow in the presence of oxygen. There are, however, varying degrees of sensitivity to oxygen. Some bacteria, such as the methanogens that produce methane gas in swamps, are killed by a few molecules of O_2, while others, such as *Clostridium,* usually can survive in O_2 but cannot grow until conditions become anaerobic.

3. *Facultative anaerobes* Organisms that can grow either in the presence or absence of oxygen. (Usually they are simply called *facultative.*) *Escherichia coli,* a member of the intestinal flora, is facultative.

There are also several other categories of organisms, such as **microaerophilic** organisms, which do best in reduced amounts of oxygen, and those that prefer more CO_2 than the amount normally found in the atmosphere. In this exercise, however, we examine the oxygen requirements of an obligate aerobe, an obligate anaerobe, and a facultative anaerobe. Try to identify them by growing each culture **aerobically** on a slant and **anaerobically** in an **agar deep**.

Definitions

Aerobic. In the presence of oxygen. Air contains about 20% oxygen.

Agar deep. A test tube filled almost to the top with agar.

Anaerobic. In the absence of oxygen.

Microaerophilic. Reduced amounts of oxygen.

Objectives

1. To understand how microbes differ in their oxygen requirements.
2. To learn to distinguish between aerobes, anaerobes, and facultative anaerobes.

Reference

Nester et al. *Microbiology: A human perspective,* 6th ed. The McGraw-Hill Companies, 2009. Chapter 4, Section 4.2.

Materials

Per team

TS broth cultures labeled A, B, and C of the following:

Escherichia coli

Micrococcus

Clostridium

TS + 0.5% glucose agar deeps, 3

TS agar slants, 3

Procedure

First Session

1. Put the agar deeps in a beaker or can filled with water to the height of the agar. Boil the

Figure 9.1 Boiling agar deeps for 10 minutes to drive out dissolved oxygen.

deeps for 10 minutes. This will not only melt the agar but also drive out all the dissolved oxygen (figure 9.1).

After the agar hardens, air gradually diffuses into the tube so that approximately the top several millimeters of the agar becomes aerobic, but the remainder of the tube is anaerobic.

2. Cool the agar in a 50°C water bath for about 10 minutes (check the temperature of the water with a thermometer). Be careful that the agar does not cool much lower than 50°C

Figure 9.2 Inoculating (a) a melted agar deep and (b) an agar slant.

(a) An agar deep   (b) Inoculating an agar slant

or it will solidify and you will have to boil it to melt it again. The tube will feel hot, but you will be able to hold it.

3. Label the tubes and slants.
4. Inoculate a melted agar deep with a loopful of culture A (figure 9.2a) and mix by rolling between the hands. Permit the agar to harden. This technique is often called a shake tube.
5. Inoculate a slant with a loopful of culture A by placing a loopful of broth on the bottom of the slant and making a wiggly line on the surface to the top of the slant (figure 9.2b).
6. Repeat with cultures B and C.
7. Incubate all slants and deeps at least 48 hours at 25°C. Some cultures grow so vigorously at 37°C that the gas produced blows apart the agar.

Second Session

1. Observe the surface of the slants and of the deeps, and record the growth. Compare to figure 9.3.

   Note: Sometimes the anaerobes seep down and grow between the agar slant and the walls of the glass tube where conditions are anaerobic, but not on the surface of the slant, which is aerobic.

2. Determine which cultures are the aerobes, the facultative anaerobes, and the anaerobes.

Figure 9.3 The appearance of aerobic and anaerobic growth in shake tubes.

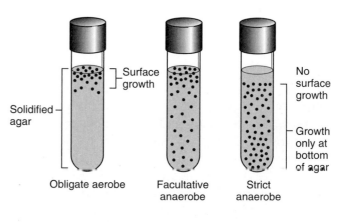

Obligate aerobe   Facultative anaerobe   Strict anaerobe

EXERCISE

9

Laboratory Report: Aerobic and Anaerobic Growth

Results

Culture	A	B	C
Appearance of growth on slant			
Appearance and location of growth in deep			

Identity of culture A
        culture B
        culture C			

Questions

1. Would you expect an obligate anaerobe to grow on a slant incubated aerobically? Why?

2. Which kind of organism would you expect to grow both in the agar deep and on the slant?

3. Which kind of organisms can grow aerobically on slants?

4. Which of the three organisms you inoculated could grow throughout the agar deep and on top? (genus)

5. Why did you boil the agar deeps longer than it took to melt the agar?

6. If air can diffuse into agar and broth, how were the obligate anaerobes grown in the broth for the class? Your instructor will explain or demonstrate.

EXERCISE 10

The Effect of Incubation Temperature on Generation Time

Getting Started

Every bacterial species has an optimal temperature—the particular temperature resulting in the fastest growth. Normally, the optimal temperature for each organism reflects the temperature of its environment. Organisms associated with animals usually grow fastest at about 37°C, the average body temperature of most warm-blooded animals. Organisms can divide more slowly at temperatures below their optimum, but there is a minimum temperature below which no growth occurs. Bacteria usually are inhibited at temperatures not much higher than their optimum temperature.

The effect of temperature can be carefully measured by determining the **generation time** at different temperatures. Generation time, or doubling time, is the time it takes for one cell to divide into two cells; on a larger scale, it is the time required for the population of cells to double. The shorter the generation time, the faster the growth rate.

Generation time can only be measured when the cells are dividing at a constant rate. To understand when this occurs, it is necessary to study the growth curve of organisms inoculated into a fresh broth medium. If plate counts are made of the growing culture, it can be seen that the culture proceeds through the four phases of growth: lag, log, stationary, and death (figure 10.1).

In the lag phase, the cells synthesize the necessary enzymes and other cellular components needed for growth. The cells then grow as rapidly as the conditions permit in the log phase, and when there are no longer sufficient nutrients or toxic products accumulate, the cells go into the stationary phase. This is followed by the death phase. The cells are growing at a constant maximum rate for the particular environment only in the log phase.

In this exercise, the generation time of *E. coli* will be compared when growing at two different temperatures. The growth of the cells can be

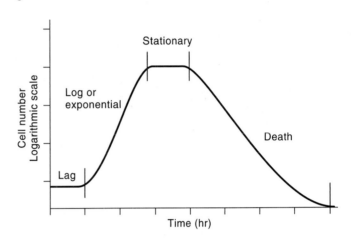

Figure 10.1 Growth curve showing the four phases of growth.

measured with a spectrophotometer or a Klett **colorimeter** because the number of cells in the culture is directly proportional to the absorbance (figure 10.2). That means that the **absorbance** (also called **optical density**) increases proportionately as the culture becomes increasingly more turbid from the multiplication of the bacteria. Readings of the incubating cultures are taken every 20 minutes for 80 minutes. The results are then plotted, and the generation time is determined.

Definitions

Absorbance. A measure of turbidity.

Generation time. The time it takes for a population of cells to double.

Colorimeter. An instrument used to measure the turbidity of bacterial growth.

Optical density (O.D.). An older but still widely used term for absorbance when used for measuring bacterial growth.

Figure 10.2 *(a)* Spectrophotometer. *(b)* Klett colorimeter. *(a)* Courtesy of Dr. Harold J. Benson. *(b)* Courtesy of VWR Scientific Company.

(a)

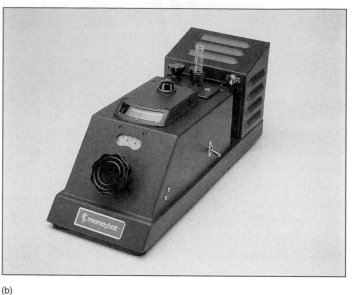

(b)

Objectives

1. To understand the phases of a growth curve.
2. To understand the effect of temperature on generation time.
3. To learn to calculate generation time.
4. To learn to use semi-log paper.

Reference

Nester et al. *Microbiology: A human perspective*, 6th ed. The McGraw-Hill Companies, 2009. Chapter 4, Section 4.8.

Materials

Per team

Culture of the following:

Escherichia coli (TS broth cultures in log phase)

TS broth, in a tube that can be read in a spectrophotometer or Klett colorimeter and that has been prewarmed in a water bath, 1

Water bath at 30°C (to be used by half the class)

Water bath at 37°C (to be used by the other half of the class)

Procedure

1. Add 0.5 ml–1.0 ml of an *E. coli* culture growing in log phase to 5.0 ml TS broth. The dilution is not important as long as the broth is turbid enough to be read at the low end of the scale (0.1 O.D. or a Klett reading of about 50). If you start with an O.D. that is too high, your last readings will reach the part of the scale that is not accurate (an O.D. of about 0.4 or about 200 on the Klett).
2. With a wavelength of 420, set the spectrophotometer at zero with an uninoculated tube of TS broth (termed a blank). Your instructor will give specific directions.
3. Take a reading of the culture and record it as 0 time. Return the tube to the assigned water bath as quickly as possible because cooling slows the growth of the organisms. Wipe off water and fingerprints from the tubes before taking a reading.
4. Read the O.D. of the culture about every 20 minutes for about 80 minutes. Record the exact time of the reading so the data can be plotted correctly.
5. Record your data—the time and O.D. readings—in your manual and on the blackboard.

6. Plot the data on semi-log graph paper (page 73). Semi-log paper is designed to convert numbers or data to $\log_{10}$ as they are plotted on the y axis. The same results would be obtained by plotting the $\log_{10}$ of each of the data points on regular graph paper, but semi-log paper simplifies this by permitting you to plot raw data and obtain the same line. Time is plotted on the horizontal x axis. Draw a straight bestfit line through the data points. The cells are growing logarithmically; therefore, the data should generate a straight line on semi-log paper (figure 10.3).

7. Also plot the data from the other temperature setting by averaging the class data on the blackboard.

8. Calculate the generation time for *E. coli* at each temperature. This can be done by arbitrarily selecting a point on the line and noting the O.D. Find the point on the line where this number has doubled. The time between these two points is the generation time.

Figure 10.3 Growth curve of cells growing in log phase at 37°C.

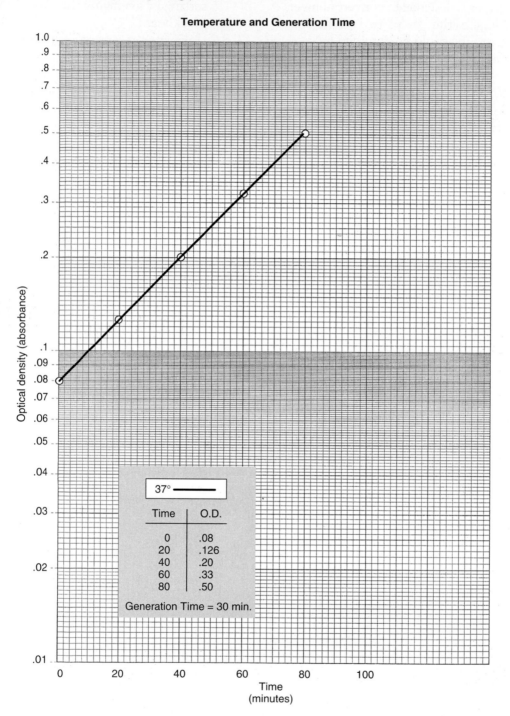

Temperature and Generation Time

Optical density (absorbance)

Time
(minutes)

37°	

Time	O.D.
0	.08
20	.126
40	.20
60	.33
80	.50

Generation Time = 30 min.

EXERCISE

10

Laboratory Report: The Effect
of Incubation Temperature on Generation Time

Results

Data: Your Temperature _____   Class Average Temperature _____

	Time	Reading	Time	Reading
1				
2				
3				
4				
5				

Generation Time

E. coli at 37°C

E. coli at 30°C

Questions

1. What is the generation time of an organism?

2. Why is it important to keep the culture at the correct incubation temperature when measuring the generation time?

3. Why is it important to use cells in log phase?

4. If the growth of two cultures, one slower than the other, were plotted on semi-log paper, which would have the steeper slope?

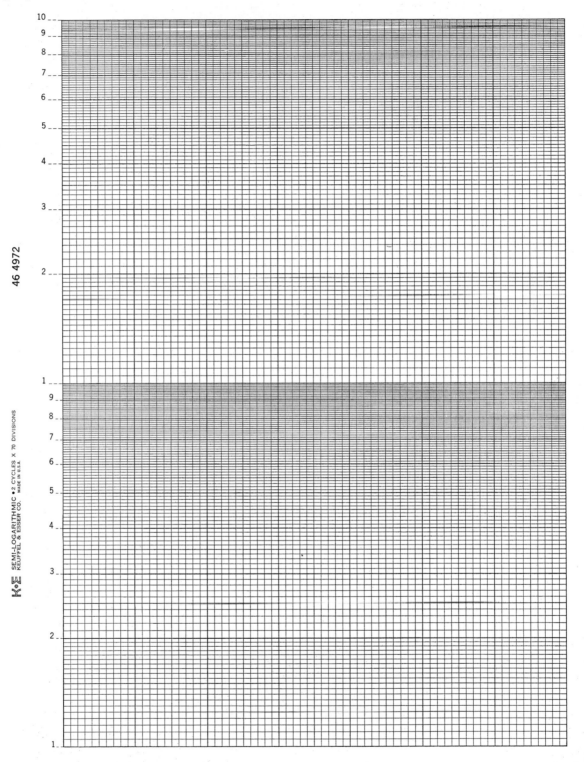

46 4972

K+E SEMI-LOGARITHMIC • 2 CYCLES X 70 DIVISIONS
KEUFFEL & ESSER CO.   MADE IN U.S.A.

NOTES:

*I*NTRODUCTION to Control of Microbial Growth

For many microbiologists, control of microbial growth means *maximization* of microbial growth—for example, when producing baker's yeast or in the production of antibiotics. To others, such as physicians and allied members of the medical profession, control means *minimization* of microbial growth—for example, the use of heat and ultraviolet light to destroy microorganisms present in growth media, gloves, and clothing. It can also imply the use of **antiseptics, disinfectants,** and **antibiotics** to inhibit or destroy microorganisms present on external or internal body parts.

Historically, Louis Pasteur (1822–1895) contributed to both areas. In his early research, he discovered that beer and wine making entailed a fermentation process involving initial growth of yeast. Later, he demonstrated that a sterile broth infusion in a swan-necked flask showed no turbidity due to microbial growth (figure I.5.1), and that upon tilting the flask, the sterile infusion became

readily contaminated. The swan-necked flask experiment was both classical and monumental in that it helped resolve a debate of more than 150 years over the possible origin of microorganisms by **spontaneous generation** (abiogenesis).

The debate was finally squelched by John Tyndall, a physicist, who was able to establish an important fact overlooked by Pasteur—namely, that some bacteria in hay infusions existed in two forms: a **vegetative form,** readily susceptible to death by boiling of the hay infusion, and a resting form, now known as an **endospore,** which could survive boiling. With this knowledge, Tyndall developed a physical method of sterilization that we now describe as **tyndallization,** whereby both vegetative cells and endospores are destroyed when the infusion is boiled intermittently with periods of cooling. For sterilization of some materials by tyndallization, temperatures below boiling are possible. Tyndallization, although a somewhat lengthy sterilization

---

Figure I.5.1 Pasteur's experiment with the swan-necked flask. If the flask remains upright, no microbial growth occurs (*1–3*). If microorganisms trapped in the neck reach the sterile liquid, they grow (*4 and 5*).

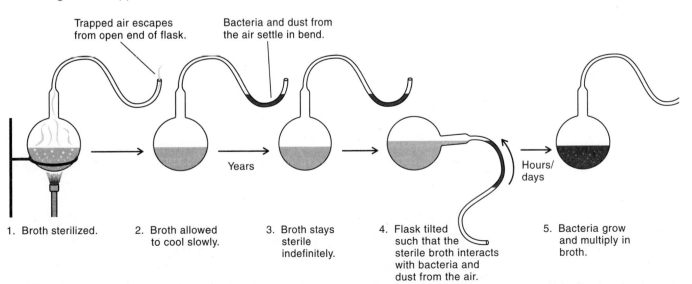

Trapped air escapes from open end of flask.

Bacteria and dust from the air settle in bend.

Years

Hours/ days

1. Broth sterilized.

2. Broth allowed to cool slowly.

3. Broth stays sterile indefinitely.

4. Flask tilted such that the sterile broth interacts with bacteria and dust from the air.

5. Bacteria grow and multiply in broth.

Introduction to Control of Microbial Growth   I–1

method, is sometimes used to sterilize chemical nutrients subject to decomposition by the higher temperatures of autoclaving.

At about this same time, chemical disinfectants that aided in healing compound bone fractures were introduced by John Lister, an English surgeon, who was also impressed with Pasteur's findings. Lister had heard that carbolic acid (phenol) had remarkable effects when used to treat sewage in Carlisle; it not only prevented odors from farmlands irrigated with sewage, but it also destroyed intestinal parasites that usually infect cattle fed on such pastures.

The control of microbial growth has many applications today, both in microbiology and in such areas as plant and mammalian cell culture. Traditional examples of it include pure culture isolation and preparation of sterile culture media, bandages, and instruments. It also includes commercial preparation of various microbial products, such as antibiotics, fermented beverages, and food.

Exercises 7, 9, and 10 touched on maximization of microbial growth. In this section, the exercises deal with minimization or elimination of microbial growth by heat, ultraviolet light, osmotic pressure, antiseptics, and antibiotics.

. . . we are too much accustomed to attribute to a single cause that which is the product of several, and the majority of our controversies come from that.

Von Liebig

EXERCISE *11*

Moist and Dry Heat Sterilization: Thermal Death Point and Thermal Death Time

Getting Started

Physical methods used in the hospital environment to control microbial growth include heat and ultraviolet light to kill microorganisms; filtration to remove microorganisms from growth media containing heat-labile substances such as enzymes as well as from air in operating rooms and cell transfer rooms; and the use of sterile gloves, masks, and clothing in these rooms to help control air convection of microorganisms present on skin and hair.

In this exercise, some effects of heat sterilization are studied since heat is commonly used to sterilize many of the materials used in hospitals and laboratories. When heat is applied, most microbes are killed, whereas when cold temperatures are applied, inhibition of microbial growth is more likely to occur. The sensitivity of a microorganism to heat is affected by its environment and genetics. Environment includes such factors as incubation temperature, chemical composition of the growth medium, and the age and concentration of cells in the growth medium. Genetically, some microorganisms are more tolerant of heat than others. Examples include the ability of microorganisms classified as **thermophiles** to grow at higher temperatures than others, and the ability of some microorganisms to produce heat-resistant structures called endospores. Members of the genera *Bacillus* and *Clostridium* are capable of endospore production.

Heat is applied in either a dry or moist form. Examples of dry heat include hot air ovens, used largely for sterilization of glassware such as petri dishes and pipets, and microincinerators, used for sterilizing needles and loops. Dry heat kills by dehydrating microorganisms, which results in **irreversible denaturation** of essential enzyme systems. Sterilization with dry heat requires considerably more time and higher temperatures than with moist heat, because dry heat penetrates to the inside of microbial cells more slowly than moist heat. Typical sterilization times and temperatures are 2 hours at 165°C for dry heat

compared to 15 minutes at 121°C at 15 psi (pounds per square inch of pressure) for moist heat. The mode of action is the same for both.

Boiling for 10 minutes is used to rid solutions such as drinking water of vegetative forms of pathogenic bacteria and other pathogens such as parasitic worms and protozoa.

Autoclaving is the most commonly used method of moist heat sterilization. Some other moist heat methods are **pasteurization, boiling,** and **tyndallization.** With autoclaving and tyndallization, both the vegetative and endospore forms of microorganisms are killed, whereas with pasteurization and boiling, usually only vegetative cells are killed.

Pasteurization, which is named for Louis Pasteur, is a moist heat process used to kill **pathogenic** bacteria and reduce the number of nonpathogenic bacteria such as **thermoduric** bacteria in such beverages as milk, beer, and wine. The beverages are heated under controlled conditions of temperature and time, either 63°C for 30 minutes or 72°C for 15 seconds. (The lower temperatures of pasteurization help preserve food flavor.) However, many endospore-forming bacteria survive pasteurization.

Tyndallization, named after the physicist John Tyndall, is sometimes used to sterilize nutrient media subject to inactivation by the higher temperatures of conventional autoclaving. It is also useful in emergencies, such as when an autoclave becomes inoperative. Tyndallization is a lengthy process requiring three days. The solution to be sterilized is usually steamed for 30 minutes in flowing steam (100°C) on each of three successive days. Between steaming times, the solution is left at room temperature. In principle, the first boiling kills all vegetative cells, the second boiling destroys newly germinated endospores, and the third boiling serves as added insurance that no living cells remain in the solution.

Two methods for determining the heat sensitivity of a microorganism are the **thermal death point** (TDP) and the **thermal death time** (TDT). The TDP is defined as the lowest temperature necessary to kill all of the microorganisms present in a culture in 10 minutes. The TDT is defined as the minimal time necessary to kill all of the microorganisms present in a culture held at a given temperature. These general principles are commonly used when establishing sterility requirements for various processes. Examples include milk, food preservation, and hospital supplies.

Definitions

Antibiotic. A chemical substance produced by a microorganism (a bacterium or a fungus) that has the capacity to inhibit the growth of or kill disease-producing microorganisms.

Antiseptic. A substance that prevents or arrests the growth or action of microorganisms, either by inhibiting their growth or by destroying them.

Autoclave. A form of moist heat sterilization, conventionally performed at 121°C for 15 minutes.

Boiling. Moist heat treatment that kills most pathogens. Conventionally, it is performed at 100°C; the time varies, although it is often done for 30 minutes.

Disinfectant. An agent that frees from infection, *e.g.*, a chemical that destroys vegetative cells but ordinarily not bacterial endospores.

Dry heat oven. Dry heat sterilization, conventionally done at 160°C–170°C for 2–3 hours.

Irreversible denaturation. A change in physical form of certain chemicals, which when heated, permanently destroys them. Enzymes constitute one such example.

Pasteurization. The use of moist heat at a temperature sufficiently high to kill pathogens but not necessarily all organisms. It is commonly used for milk products.

Pathogen. An organism able to cause disease.

Spontaneous generation. A discredited idea that living organisms can arise from non-living material.

Thermal death point (TDP). The lowest temperature at which all microbes in a culture are killed after a given time.

Thermal death time (TDT). The minimal time necessary to kill all microbes in a culture held at a given temperature.

Thermoduric bacteria. Microbes able to survive conventional pasteurization, usually 63°C for 30 minutes or 72°C for 15 seconds.

Thermophile. An organism able to grow at temperatures above 55°C.

Tyndallization. The process of using repeated cycles of heating and incubation to kill spore-forming bacteria.

Vegetative form. Cells growing and dividing in typical fashion and not forming endospores.

Objectives

1. To provide background information about physical sterilization methods requiring either moist heat or dry heat. Included in the discussion of moist heat methods are autoclaving, boiling, tyndallization, and pasteurization.
2. To introduce a quantitative laboratory method for determining the susceptibility of different bacteria to the lethal effect of moist heat—thermal death point (TDP) and thermal death time (TDT).
3. To demonstrate laboratory equipment commonly used for physical sterilization of moist and dry materials: the steam autoclave and the **dry heat oven.**

References

Block, S. S. (ed). *Disinfection, sterilization and preservation*, 4th ed. Philadelphia, PA: Lea & Febiger, 1991.

Frazier, W. C., and Westhoff, D. *Food microbiology*, revised. New York: McGraw-Hill Book Company, 1988. Contains relevant information on principles of food preservation and spoilage.

Nester et al. *Microbiology: A human perspective*, 6th ed. The McGraw-Hill Companies, 2009. Chapter 5, Section 5.3.

Materials

Cultures of the following:

> 24-hour 37°C *Escherichia coli* cultures in 5 ml TS broth
>
> Spore suspension in 5 ml of sterile distilled water of a 4–5 day 37°C nutrient agar slant culture of *Bacillus subtilis*

5 ml of TS broth, 5 tubes per student

Two large beakers or cans, one for use as a water bath and the other for use as a reservoir of boiling water (per student)

Either a ring stand with wire screen and a Bunsen burner, a hot plate, or a heat block (per student)

One thermometer (per student)

A community water bath

A vortex apparatus (if available)

Procedure

First Session: Determination of Thermal Death Point and Thermal Death Time

Note: A procedural culture and heating distribution scheme for 8 students is shown in table 11.1. For this scheme, each student receives one broth culture, either *Escherichia coli* or *Bacillus subtilis*, to be heated at one of the assigned temperatures (40°C, 55°C, 80°C, or 100°C).

Note: As an alternative, instead of each student preparing his or her own water bath, the instructor can provide four community water baths, preset at 40°C, 55°C, 80°C, and 100°C.

1. Suspend your assigned culture by gently rolling the tube between your hands, followed by aseptically transferring a loopful to a fresh tube of broth (label "0 time Control").

2. Fill the beaker to be used as a water bath approximately half full with water, sufficient to totally immerse the broth culture without dampening the test tube cap.

3. Place your tube of broth culture in the water bath along with an open tube of uninoculated broth in which a thermometer has been inserted for monitoring the water bath temperature.

4. Place the water bath and contents on either the ring stand or hot plate and heat almost to the assigned temperature. One or two degrees before the assigned temperature is reached, remove the water bath from the heat source and place it on your bench top. The temperature of the water bath can now be maintained by periodically stirring in small amounts of boiling water obtained from the community water bath.

   Note: Students with the 100°C assignment may wish to keep their water bath on the heat source, providing the water can be controlled at a low boil.

5. After 10 minutes of heating, resuspend the broth culture either by vortexing or by gently tapping the outside of the tube. Aseptically transfer a loopful to a fresh tube of broth (label "10 minutes").

6. Repeat step 5 after 20, 30, and 40 minutes.

7. Write the initials of your culture as well your initials on all 5 tubes and incubate them in the 37°C incubator for 48 hours.

Demonstration of the Steam-Jacketed Autoclave and Dry Heat Oven

The Steam-Jacketed Autoclave

Note: As the instructor demonstrates the special features of the autoclave, follow the diagram in figure 11.1. Note the various control valves—their

Table 11.1 Culture and Heating Temperature Assignments (8 Students)

Bacterial Culture Assignment	Water Bath Temperature Assignment			
	40°C	55°C	80°C	100°C
Escherichia coli	1	2	3	4
Bacillus subtilis	5	6	7	8

Figure 11.1 Steam-jacketed autoclave. Entering steam displaces air down and out through a port in the bottom of the chamber. Dry objects are placed in the autoclave in a position to avoid trapping air. Watery liquids generate their own steam.

function and method of adjustment (exhaust valve, chamber valve, and safety valve); the steam pressure gauge; the thermometer and its location; and the door to the chamber. Also make a note of the following special precautions necessary for proper sterilization.

1. Material sensitivity. Certain types of materials, such as talcum powder, oil, or petroleum jelly, cannot be steam sterilized because they are water repellent. Instead, dry heat is used. Some materials are destroyed or changed by the standard autoclave temperature of 121°C; for example, some sugars are hydrolyzed and some medications are chemically changed. In such instances, the autoclave may be operated at a lower pressure and temperature for a longer period of time. A heat-sensitive fluid material can normally be sterilized by filtration. Filtration removes the bacteria. Keep in mind that the smaller the filter pores, the slower the rate of filtration.

2. Proper preparation of materials. The steam must be in direct contact with all materials to be sterilized. Therefore, media container closures such as metal caps with air passages, loosened screw cap lids, aluminum foil (heavy grade), and sometimes nonabsorbant cotton plugs are used.

3. Proper loading of supplies. There must be ample space between packs and containers so that the steam can circulate. When using cotton plugs, they should be loosely covered with foil to prevent moisture from the autoclave condensing on them during cooling.

4. Complete evacuation of air from the chamber. This is necessary before replacement with steam. Older models may require manual clearance, but in newer models it is automatic.

5. Proper temperature. Autoclaving at a pressure of 15 psi achieves a temperature of 121°C (250°F) at sea level. If the temperature gauge

does not register this reading, trapping of colder air in the autoclave is indicated, which lowers the temperature.

Note: The temperature is critical, not the pressure.

6. Adequate sterilization time. After the chamber temperature reaches 121°C, additional time is required for the heat to penetrate the material. The larger the size of individual containers and packs, the more time required. Time must be adjusted to the individual load size. Also, highly contaminated materials require more time.

7. Completion of the autoclaving process. Rapid reduction in steam pressure can cause fluids to boil vigorously through the caps or to explode. Drop steam pressure gradually as cooling occurs and, when possible, allow material to dry in the autoclave. If removed while moist, the wrappings or plugs may provide a means for reentry of bacteria present in the room air.

The Dry Heat Oven

The hot air, or dry heat, oven is used in most laboratories for both drying glassware and sterilization.

When using the dry heat oven, the following guidelines are important:

1. Types of material suitable for sterilization include oil, petroleum jelly, Vaseline, metal containers, and dry, clean glassware.

2. An oven with circulating air takes about half the sterilization time of a static air oven. Better heat transfer occurs with circulating air.

3. Proper packaging is necessary to assure air circulation to the inside surfaces. For example, syringes must be separate from the plunger so that all surfaces are exposed to circulating air.

4. Sterilization of dirty materials should be avoided. The presence of extraneous materials such as protein delays the process and may allow bacteria to survive inside the material.

5. In part 2 of the Laboratory Report, prepare a list of materials for your class that are sterilized in the autoclave. For each one, indicate the standard temperature, pressure, and time used for sterilization.

   Do the same for materials sterilized in the dry heat oven. Indicate the oven sterilization temperature, time, and reasons for sterilizing there.

Second Session

After 48 hours, examine your broth tubes for the presence or absence of turbidity (growth). Write your results in the appropriate place in the table on the blackboard of your classroom. When all the results are entered, transfer them to table 11.2 of the Laboratory Report.

NOTES:

EXERCISE

11

Laboratory Report: Moist and Dry Heat Sterilization: Thermal Death Point and Thermal Death Time

Results

1. Determination of thermal death point and thermal death time:

Table 11.2   Bacterial Growth at Assigned Temperatures and Times

Culture	40°C					55°C					80°C					100°C				
	C	10	20	30	40	C	10	20	30	40	C	10	20	30	40	C	10	20	30	40
Escherichia coli																				
Bacillus subtilis																				

Note: C = Control; 10, 20, 30, 40 = minutes of heating the inoculated culture at the assigned temperature; Use a + sign for growth and a – sign for no growth

a. Determine the thermal death time for each culture.

   Thermal Death Time (Minutes)

   Escherichia coli   ____

   Bacillus subtilis   ____

b. Determine the thermal death point for each culture.

   Thermal Death Point (°C)

   Escherichia coli   ____

   Bacillus subtilis   ____

2. Evaluation of materials sterilized in our laboratory with moist heat (autoclave) and dry heat (hot air oven):

    a. List of materials sterilized with the autoclave (see Procedure for criteria):

    b. List of materials sterilized with the dry heat oven (see Procedure for criteria):

Questions

1. Discuss similarities and differences between determining thermal death point and thermal death time.

2. How would you set up an experiment to determine to the minute the TDT of *E. coli*? Begin with the data you have already collected.

3. A practical question related to thermal death time (TDT) relates to a serious outbreak of *E. coli* infection in early 1993 when people ate insufficiently grilled hamburgers. How would you set up an experiment to determine the TDT of a solid such as a hamburger? Assume that the thermal death point (TDP) is 67.2°C (157°F), the temperature required by many states to cook hamburger on an open grill. What factors would you consider in setting up such an experiment?

4. What is the most expedient method for sterilizing a heat-sensitive liquid that contains a spore-forming bacterium?

5. List one or more materials that are best sterilized by the following processes:

    a. Membrane filtration

    b. Ultraviolet light

    c. Dry heat

    d. Moist heat

    e. Tyndallization

6. What are three advantages of using metal caps rather than cotton for test tube closures? Are there any disadvantages?

7. How would you sterilize a heat-sensitive growth medium containing thermoduric bacteria?

8. Was the *Bacillus subtilis* culture sterilized after 40 minutes of boiling? If not, what is necessary to **assure** sterility by boiling?

EXERCISE 12

Control of Microbial Growth with Ultraviolet Light

Getting Started

Ultraviolet (UV) light is the component of sunlight that is responsible for sunburn. It can also kill microorganisms by acting on their DNA, causing mutations. It consists of very short **nonionizing wavelengths of radiation** (175–350 nm) located just below blue light (450–500 nm) in the **visible spectrum** (figure 12.1).

The actual mechanism of mutation is the formation of **thymine dimers** (figure 12.2). Two adjacent **thymines** on a DNA strand bind to each other; when the DNA is replicated, an incorrect base pair is frequently incorporated into the newly synthesized strand. This may cause mutation, and if there is sufficient radiation, ultimately, the death of the cell.

UV light does not penetrate surfaces and will not go through ordinary plastic or glass. It is only useful for killing organisms *on* surfaces and in the air. Sometimes UV lights are turned on in operating rooms and other places where airborne bacterial contamination is a problem. Since UV light quickly damages the eyes, these lights are turned on only when no one is in the irradiated area.

Bacteria vary in their sensitivity to UV light. In this exercise, the sensitivity of *Bacillus* endospores

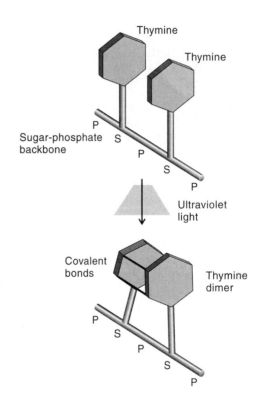

Figure 12.2 Thymine dimer formation. Covalent bonds form between adjacent thymine molecules on the same strand of DNA. This distorts the shape of the DNA and prevents replication of the changed DNA.

Figure 12.1 Germicidal activity of radiant energy.

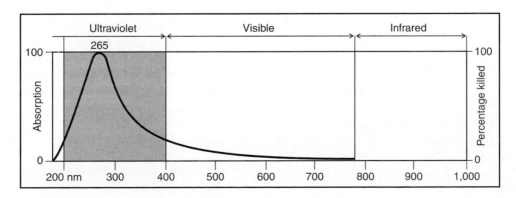

will be compared with non-spore-forming cells. You will also irradiate a mixed culture, such as organisms in soil or hamburger, to compare the resistance of different organisms. Immediately following irradiation, it is important to place the samples in a dark container since some cells potentially killed by UV light recover when exposed to longer wavelengths of light (light repair).

Definitions

Light repair.   DNA repair in cells previously exposed to UV light by a DNA repair enzyme that requires visible light.

Nonionizing wavelengths of radiation.   Noncharged wavelengths of radiation between 175 nm and 350 nm. This light is invisible and is exemplified by UV light. Beta rays and X rays are examples of ionizing wavelengths of radiation.

Thymine dimer.   A molecule formed when two adjacent thymine molecules in the same strand of DNA covalently bond to one another.

Thymine.   One of the four nucleotide subunits of DNA.

Visible spectrum.   Uncharged wavelengths of radiation between 400 and 800 nm.

Reference

Nester et al. *Microbiology: A human perspective*, 6th ed. The McGraw-Hill Companies, 2009. Chapter 5, Section 5.4.

Materials

Cultures of the following:

Suspension of *Bacillus* spores in sterile saline

E. coli in TS broth

Raw hamburger or soil mixed with sterile water

TS agar plates, 3 per team

UV lamp with shielding. An 18- to 36-inch fluorescent bulb is ideal. It enables uniform exposure of 3–6 partially opened petri dishes.

Sterile swabs, 3

Safety glasses for use with the UV lamp

Dark box for storing plates after UV exposure

Procedure

First Session

Safety Precautions: (1) The area for UV irradiation should be in an isolated part of the laboratory. (2) Wear safety glasses as a precautionary measure when working in this area. (3) Never look at the UV light after turning it on since it could result in severe eye damage. Skin damage is also a slight possibility.

1. Dip a sterile swab in a suspension of *Bacillus* spores and swab an agar plate in three directions as shown in figure 14.3.
2. Repeat the procedure with a suspended *E. coli* broth culture.
3. For the third plate, you can either dip the swab into a mixture of sterile water and hamburger or sterile water and soil.
4. Place the plates under a UV lamp propped up about 20 cm from the bench surface. Open the petri dish(es) and partially cover each plate with the lid. The part of the plate protected by the lid will be the control because UV radiation does not penetrate most plastic.
5. Put on your safety glasses and turn on the UV light. Expose the plates to UV radiation for 3 minutes. (Your instructor may assign different exposure times to some teams.)
6. Turn off the UV light. Cover the plates, invert them, and place them in a covered container. Incubate at 37°C for 48 hours.

Second Session

Observe the plates and record your findings in the Laboratory Report.

EXERCISE

12

Laboratory Report: Control of Microbial Growth
with Ultraviolet Light

Results

1. Record your observations for control and treated sides of petri dishes exposed to UV light. Record your
    results on the blackboard.

	Assigned Exposure	Optional Class Exposures
Bacillus endospores		
Escherichia coli		
Soil or meat suspension		

Score irradiated growth:
++++   no different than nonirradiated control
 +++   slightly different than nonirradiated control
  ++   significantly less growth than nonirradiated control
   +   less than 10 colonies

    a. Plate containing *Bacillus* spores

    b. Plate containing *E. coli*

    c. Plate containing either a raw hamburger suspension or soil suspension (indicate which one
    you used)

2. Make a drawing of each plate.

3. Which organisms were most resistant to UV light? _____ least resistant? _____

Questions

1. Why can't you use UV light to sterilize microbiological media, *e.g.*, agar or broth?

2. How does UV light cause mutations?

3. What is the result of increasing the time of exposure to UV light?

4. Give a possible reason some organisms in the soil (or meat) were able to grow after exposure to UV light but others were not.

5. Frequently organisms isolated from the environment are pigmented, while organisms isolated from the intestine or other protected places are not. Can you provide an explanation for this?

6. Mutations can lead to cancer in animals. Explain why persons living in the southern half of the United States have a higher incidence of skin cancer than those in the northern half.

NOTES:

EXERCISE

13

Osmotic Pressure and Its Effect on the Rate and Amount of Microbial Growth

Getting Started

Osmosis, which is derived from the Greek word "alter," refers to the process of flow or diffusion that takes place through a **semipermeable membrane.** In a living cell, the cytoplasmic membrane, located adjacent to the inside of the cell wall, represents such a membrane (figure 13.1*a*).

Both the cytoplasmic membrane and the cell wall help prevent the cell from either bursting (**plasmoptysis**) (figure 13.1*a*) or collapsing (**plasmolysis**) (figure 13.1*b*) due to either entry or removal of water from the cell, respectively. The **solute** concentration both inside and outside the cell determines which, if any, process happens. When the solute concentration inside the cell is the same as the solute concentration on the outside of the cell (**isotonic**), the cell remains intact. When the solute concentration outside the cell is less than the solute concentration inside the cell, an inward **osmotic pressure** occurs, and water enters the cell in an attempt to equalize the solute concentration on either side of the cytoplasmic membrane. If the solute concentration outside the cell is sufficiently low (**hypotonic**), the cell will absorb water and sometimes burst (plasmoptysis). However, this rarely occurs due to the rigidity of the cell wall. The reverse phenomenon, cell shrinkage followed by cell lysis (plasmolysis), can occur when the cell is placed in a more concentrated (**hypertonic**) solution. This can become a life-and-death problem if too much water is removed from the cells (figure 13.1*b*).

When placed in an isotonic solution, some cells recover, although there are many genera that die once the external osmotic pressure exceeds their limitations. This concept is the basis of food preservation—the use of high salt concentrations (for cheese and pickle brine) and high sugar concentrations (in honey and jams).

In general, fungi (yeasts and molds) are much more resistant to high external solute concentrations than are bacteria, which is one reason fungi

Figure 13.1 Movement of water into and out of cells. *(a)* Low and *(b)* very high salt-containing solutions. The cytoplasmic membrane is semipermeable and only allows water molecules to pass through freely.

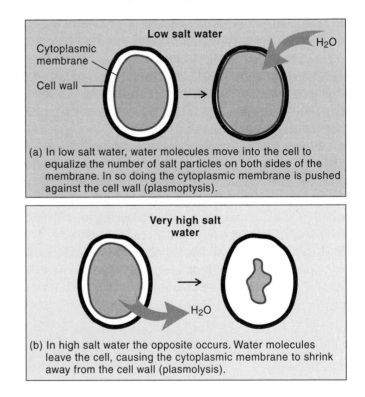

(a) In low salt water, water molecules move into the cell to equalize the number of salt particles on both sides of the membrane. In so doing the cytoplasmic membrane is pushed against the cell wall (plasmoptysis).

(b) In high salt water the opposite occurs. Water molecules leave the cell, causing the cytoplasmic membrane to shrink away from the cell wall (plasmolysis).

can grow in or on jelly, cheese, and fruit. There are, of course, exceptions among the bacteria; for example, the genus *Halobacterium* in the *Archaea* is found in nature growing in water with a high salt content, like Great Salt Lake in Utah; and the genus *Micrococcus halobius* in the family *Micrococcaceae* is sometimes found in nature growing on highly salted (25–32%) protein products, such as fish and animal hides. It is interesting to note that all of these bacteria produce a red pigment.

There are also some salt-tolerant *Staphylococcus* strains able to grow at salt concentrations greater than 10% (w/v), enabling them to grow on skin surfaces. The salt-loving (**halophilic**) bacteria are also unique in that, like all *Archeae*, they lack muramic acid as a bonding agent in their cell walls. Instead, their cell walls are believed to contain sodium and potassium ions. These ions help confer cell wall rigidity, perhaps helping explain the reason they require such high salt concentrations for good growth.

Yeasts and molds able to grow in high sugar (50–75%) and sometimes high salt (25–30%) concentrations are termed **saccharophilic** and halophilic fungi, respectively. Some of these yeast and fungus genera are *Debaromyces*, *Saccharomyces*, *Zygosaccharomyces*, *Aspergillus*, and *Penicillium*.

In this exercise, you will examine the ability of some of the previously mentioned halophiles and saccharophiles to grow on the surface of TYEG (Trypticase yeast extract glucose) agar plates containing increasing concentrations of salt and sugar. *Escherichia coli* is added as a salt-sensitive Gram-negative rod control. You will also have an opportunity to examine any changes in cell form with increasing salt and sugar concentrations.

Keep in mind that all of these halophilic and saccharophilic microbes are characterized by an increase in lag time and a decrease in growth rate and in the amount of cell substance synthesized. In some ways, their **growth curve** (see exercise 10) parallels what happens when they are grown at a temperature below their optimal growth temperature. For example, halobacteria have a **generation time** of 7 hours and halococci have one of 15 hours.

Definitions

Generation time. The time required for one cell to divide into two cells.

Growth curve. A curve describing the four readily distinguishable phases of microbial growth: lag, log, stationary, and death.

Halophilic microbe. A salt-requiring organism able to grow in a medium containing a salt concentration high enough to inhibit other organisms.

Hypertonic fluid. A fluid having an osmotic pressure greater than another fluid with which it is compared.

Hypotonic fluid. A fluid having an osmotic pressure lower than another fluid with which it is compared.

Isotonic fluid. A fluid having the same osmotic pressure as another fluid with which it is compared.

Osmotic pressure. The pressure exerted by water on a membrane as a result of a difference in the concentration of solute molecules on each side of the membrane.

Plasmolysis. Contraction or shrinking of the cytoplasmic membrane away from the cell wall due to a loss of water from the cell.

Plasmoptysis. The bursting of protoplasm from a cell due to rupture of the cell wall when absorbing excess water from the external environment.

Saccharophilic microbe. An organism able to grow in environments containing high sugar concentrations.

Semipermeable membrane. A membrane, such as the cytoplasmic membrane of the cell, that permits passage of some materials but not others. Passage usually depends on the size of the molecule.

Solute. A dissolved substance in a solution.

Objectives

1. To provide an introduction to osmotic pressure and to show how it can be used to inhibit growth of less osmotolerant microbes while allowing more osmotolerant microbes to grow, although often at a considerably slower growth rate.
2. To show that some microorganisms either require or grow better in an environment containing high concentrations of salt (halophilic) or sugar (saccharophilic).

Reference

Nester et al. *Microbiology: A human perspective*, 6th ed. The McGraw-Hill Companies, 2009. Chapter 3, Section 3.4, Chapter 4, Section 4.5.

Materials

Cultures (may be shared by 2–4 students) of the following:

Use Trypticase Yeast Extract Glucose (TYEG) agar slants for *Escherichia coli*, *Micrococcus luteus*, and *Saccharomyces cerevisiae*

Use American Type Culture Collection (ATCC) medium 213 for the Preceptrol strain of *Halobacterium salinarium*

Incubate *E. coli*, *M. luteus*, and *S. cerevisiae* cultures for 24 hours at 35°C; incubate *H. salinarium* culture for 1 week (perhaps longer) at 35°C

TYEG agar plates containing 0.5, 5, 10, and 20% NaCl, 4 plates

TYEG agar plates containing 0, 10, 25, and 50% sucrose, 4 plates

Procedure

First Session

Note: One student of the pair can inoculate the four TS agar plates containing 0.5, 5, 10, and 20% NaCl, while the other student inoculates the four plates containing 0, 10, 25, and 50% sucrose.

1. First, use a glass-marking pen to divide the bottom of the eight plates in quadrants, and label with the initials of the four test organisms; for example, *E.c.* for *Escherichia coli*, etc. Also label the underside of each plate with the salt or sugar concentrations and your name.
2. Using aseptic technique, remove a loopful from a culture and streak the appropriate quadrant of your plate in a straight line approximately 1 inch long. Then reflame your loop, cool it for a few seconds, and make a series of cross streaks approximately 1/2 inch long in order to initiate single colonies for use in studying colonial morphology (figure 13.2). Repeat the inoculation procedure used for the first culture in the appropriate quadrant of the remaining three agar plates.

Figure 13.2
Streaking procedure for isolation of single colonies.

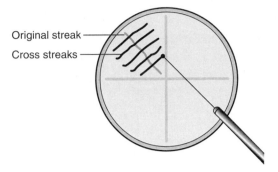

Original streak

Cross streaks

3. Repeat the inoculation procedure for the remaining three test organisms.
4. Invert and incubate the eight plates at 30°C.
5. Observe the plates periodically (up to 1 week or more if necessary) for growth.

Second Session

1. Examine your plates for the presence (+) or absence (−) of growth. For growth, use one to three + signs (+ = minimal, + + = some, and + + + = good). Enter results in table 13.1 (various salt concentrations) and table 13.2 (various sugar concentrations) of the Laboratory Report.
2. Compare the colonial growth characteristics of cultures grown on agar media containing increasing salt and sugar concentrations. Make notes of any marked changes in colony color, colony size (in mm), and colony texture: dull or glistening, rough or smooth, and flat or raised. Record your findings in the Laboratory Report.
3. Prepare wet mounts of bacteria and yeast colonies showing marked changes in visual appearance with increasing salt and sugar concentrations. Examine bacteria with the oil immersion objective lens and yeasts with the high-dry objective lens. Look for plasmolyzed cells and other changes, such as cell form and size. Prepare drawings of any such changes in the Laboratory Report.

NOTES:

EXERCISE

13

Laboratory Report: Osmotic Pressure and Its Effect on the Rate and Amount of Microbial Growth

Results

1. Examination of petri dish cultures for the presence (+) or absence (−) of growth in the presence of increasing salt (table 13.1) and sugar concentrations (table 13.2). Use a series of one to three + signs to describe the amount of growth.

Table 13.1 Presence or Absence of Growth on TYEG Agar Plates Containing NaCl and Incubated for 48 Hours–1 Week

Culture	NaCl Concentration (%)							
	0.5		5		10		20	
	48hr	1wk	48hr	1wk	48hr	1wk	48hr	1wk
Escherichia coli								
Halobacterium salinarium								
Micrococcus luteus								
Saccharomyces cerevisiae								

Table 13.2 Presence or Absence of Growth on TYEG Agar Plates Containing Sucrose and Incubated for 48 Hours–1 Week

Culture	Sucrose Concentration (%)							
	0		10		25		50	
	48hr	1wk	48hr	1wk	48hr	1wk	48hr	1wk
Escherichia coli								
Halobacterium salinarium								
Micrococcus luteus								
Saccharomyces cerevisiae								

2. Comparison of the colonial growth characteristics of cultures inoculated on agar media containing increasing amounts of salt or sugar.

   a. *Escherichia coli*

NaCl %	Growth	Colony Color	Colony Size	Colony Texture
0.5				
5				
10				
20				

Sucrose %	Growth	Colony Color	Colony Size	Colony Texture
0				
10				
25				
50				

b. *Halobacterium salinarium*

NaCl %	Growth	Colony Color	Colony Size	Colony Texture
0.5				
5				
10				
20				

Sucrose %	Growth	Colony Color	Colony Size	Colony Texture
0				
10				
25				
50				

c. *Micrococcus luteus*

NaCl %	Growth	Colony Color	Colony Size	Colony Texture
0.5				
5				
10				
20				

Sucrose %	Growth	Colony Color	Colony Size	Colony Texture
0				
10				
25				
50				

d. *Saccharomyces cerevisiae*

NaCl %	Growth	Colony Color	Colony Size	Colony Texture
0.5				
5				
10				
20				

Sucrose %				
0				
10				
25				
50				

3. Examine wet mounts of bacteria and yeast colonies showing *marked* changes in appearance from the control colonies. How do the cells appear?

Questions

1. From your studies, which organism(s) tolerate salt best? _____ least? _____
2. Which organism(s) tolerate sugar best? _____ least? _____
3. Compare bacteria and yeast with respect to salt tolerance. Bear in mind both colonial and cellular appearance in formulating your answers.

4. Compare bacteria and yeast with respect to sugar tolerance. Bear in mind both colonial and cellular appearance in formulating your answer.

5. What evidence did you find of a *nutritional requirement* for salt or sugar in the growth medium?

6. Matching
   Each answer may be used one or more times.
   a. *Halobacterium*   ___ osmosensitive
   b. *Saccharomyces*   ___ long generation time
   c. *Escherichia coli*   ___ saccharophilic
   d. *Micrococcus*   ___ osmotolerant

7. Matching
   Choose the best answer. Each answer may be used one or more times or not at all.
   a. Plasmolysis   ___ isotonic solution
   b. Plasmoptysis   ___ hypotonic solution
   c. Normal cell growth   ___ hypertonic solution
                           ___ swelling of cells

EXERCISE 14

Antiseptics and Antibiotics

Getting Started

It was during a visit through Central Europe in 1908 that I came across the fact that almost every farmhouse followed the practice of keeping a moldly loaf on one of the beams in the kitchen. When asked the reason for this I was told that this was an old custom and that when any member of the family received an injury such as a cut or bruise a thin slice from the outside of the loaf was cut off, mixed into a paste with water and applied to the wound with a bandage. I was assured that no infection would then result from such a cut.

Dr. A. E. Cliffe, a Montreal biochemist

Chemicals such as antibiotics have been in existence a long time. However, the therapeutic properties of antibiotics simply were not recognized until Alexander Fleming's discoveries in the 1930s.

By definition, **antibiotics** are chemicals produced and secreted by microorganisms, especially bacteria and fungi, that can inhibit or destroy the growth of **pathogens,** often by altering an **essential metabolic pathway.** To be effective in human medicine, ideally they should be nontoxic to the human host and should discourage the formation of microbial strains resistant to the antibiotic (see exercise 15). Many antibiotics in use today are synthesized in the chemical laboratory.

Another interesting group of synthetic chemicals are sulfa drugs, which originate from the azo group of dyes. One of these drugs, marketed under the name of Prontosil, inhibited microbial growth when tested *in vivo* (initially in mice and later in humans), but when tested ***in vitro*** (test tubes) against streptococci nothing happened. In the 1930s, scientists at the Pasteur Institute in Paris showed that if Prontosil is chemically reduced it is not

Figure 14.1 *(a)* Structures of sulfanilamide (sulfa drug) and of para-aminobenzoic acid (PABA). The portions of the molecules that differ from each other are shaded. *(b)* Reversible competitive inhibition of folic acid synthesis by sulfa drug. The higher the concentration of sulfa drug molecules relative to PABA, the more likely that the enzyme will bind to the sulfa drug, and the greater the inhibition of folic acid synthesis.

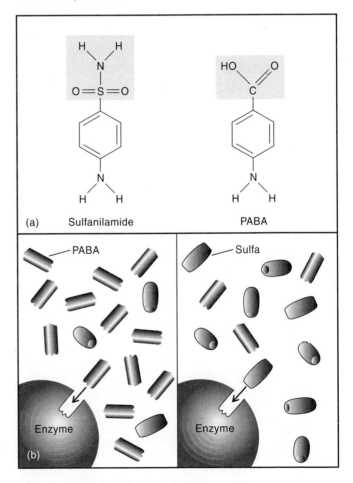

only active in the body, but it can also kill bacteria *in vitro.* Prontosil is active *in vivo* (but not *in vitro*) because enzymes within mice and humans reduce Prontosil to a smaller active molecule, known as sulfanilamide (figure 14.1).

The inhibitory action of the sulfanilamide-type drugs is termed **competitive inhibition,** in which the sulfanilamide acts as an **antimetabolite.** Antimetabolites are molecules resembling essential cellular compounds, but block cell function when incorporated into the cell's metabolism. The sulfanilamide component replaces para-aminobenzoic acid (PABA), an **essential metabolite,** as a necessary part of folic acid. Folic acid is an essential **coenzyme** involved in amino acid synthesis.

Antiseptics are chemicals able to inhibit *in vivo* **sepsis** (infection). They do not kill the sepsis-producing agent; they merely inhibit its growth. Antiseptic chemicals must be sufficiently nontoxic to allow application to skin and mucous membranes, such as the use of Listerine® for gargling.

These same chemicals act as **disinfectants** (a chemical able to kill vegetative forms but not necessarily endospore forms of bacteria) when used at higher concentration levels. Toxicity is a major factor in determining usage of a chemical as either an antiseptic or disinfectant. Most *inorganic* heavy metal compounds can be used only as disinfectants (for example, mercurials). *Organic* heavy metal mercurial compounds (like mercurochrome) and also quaternary ammonium compounds and phenolics can be used as either antiseptics or disinfectants, depending on the concentration used.

The mode of action of antiseptics and disinfectants differs from antibiotics in that they act largely by denaturing proteins. They also lower surface tension, which is accompanied by cytoplasmic membrane dissolution (quaternary ammonium compounds), and act as oxidizing agents (chlorine-containing chemicals). A list of commonly used antiseptics and disinfectants and their area of application is shown in table 14.1.

An assay still used in many clinical laboratories to test the potency of antibiotics and drugs is a filter paper discagar diffusion procedure commonly known

Table 14.1   Chemical Compounds Commonly Used in Hospitals for Controlling Growth of Microorganisms

Sodium hypochlorite (5%)	Disinfectant	External surfaces, such as tables
Iodine (1% in 70% alcohol)	Disinfectant	External surfaces, such as tables
Iodophors (70 ppm avail. I_2)	Disinfectant	External surfaces, such as tables
Quatenary ammonium compounds	Disinfectant	External surfaces, such as tables
Phenol (5%), carbolic acid, source coal tar	Disinfectant	External surfaces, such as tables
Hexachlorophene (pHisoHex®)	Disinfectant	Presurgical hand washing
Formaldehyde (4%)	Disinfectant	Oral and rectal thermometers
Iodophors (70 ppm avail. I_2)	Disinfectant	Oral and rectal thermometers
Alcohol, ethanol (70%)	Antiseptic	Skin
Iodine (tincture in alcohol with KI)	Antiseptic	Skin
Iodophors	Antiseptic	Skin
Organic mercury compounds (merthiolate, mercurochrome)	Antiseptic	Skin
Hydrogen peroxide (3%)	Antiseptic	Superficial skin infections
Potassium permanganate	Antiseptic	Urethral and superficial skin fungus infections
Silver nitrate (1%)	Antiseptic	Prevention of eye infections in newborn babies
Zinc oxide paste	Antiseptic	Diaper rash
Zinc salts of fatty acids (Desenex®)	Antiseptic	Athlete's foot
Glycerol (50%)	Antiseptic	Prevent bacterial growth in stool and surgical specimens
Ethylene oxide gas (12%)	Sterilization	Linens, syringes, etc.
Formaldehyde (20% in 70% alcohol)	Sterilization	Metal instruments
Glutaraldehyde (pH 7.5 or more)	Sterilization	Metal instruments

Table 14.2 Chart Containing Zone Diameter Interpretive Standards for Determining the Sensitivity of Bacteria to Antimicrobial Agents

Antimicrobial Agent	Disc Content	Susceptible	Intermediate	Resistant
Penicillin G	6 μg	≥26	23–25	≤22
when testing staphylococci & streptococci	6 μg	≥26	23–25	≤22
when testing enterococci	6 μg	≥12		≤11
Streptomycin	30 μg	≥26	23–25	≤22
Tetracycline	30 μg	≥26	23–25	≤22
Chloramphenicol	30 μg	≥26	23–25	≤22
Trimethoprim & Sulfamethoxazole*	1.25 + 23.75 μg	≥26	23–25	≤22
Erythromycin	15 μg	≥26	23–25	≤22

Source: Fluka (A Sigma-Aldrich Company). Antimicrobial Susceptibility Test Discs, Inhibition Zone Diameter Interpretation Application Leaflet (2003). Used by permission.

*NOTE: Sulfamethoxazole (a sulfa drug) is usually given in combination with trimethoprim.

as the Kirby-Bauer test. A filter paper discagar diffusion method is also used for determining the potency of antiseptics. In this exercise, you will have an opportunity to determine both antiseptic and antibiotic potency with a modified Kirby-Bauer test.

Discs of filter paper impregnated with antibiotic solutions in the same range of concentrations obtainable in the human body are placed on an agar plate heavily seeded with the test bacterium. When incubated, the bacteria grow in a smooth lawn of confluent growth except in a clear zone around the antibiotic disc. The clear growth inhibition zone does not necessarily indicate degree of microbial susceptibility to the antibiotic, for zone size itself does not indicate if the antibiotic is appropriate for use in clinical treatment. Some are smaller molecules and diffuse faster producing a larger zone. When conducting the precisely controlled Kirby-Bauer test, various special conditions, such as 2- to 5-hour cultures, controlled inoculum size, and short incubation periods, are necessary. Such conditions are difficult to achieve in the time frame of most classrooms. Consequently, the simpler test performed in this exercise will demonstrate the Kirby-Bauer test principle without the added restrictions. Upon observation, the diameter of the clear zone of inhibition surrounding each antibiotic disc can be compared to that contained for the antibiotic on a standard chart (table 14.2). This chart will help determine if the test bacterium is resistant, intermediate, or sensitive to the antibiotic in question. (A comparison chart is not available for antiseptics.)

Definitions

Antibiotic. A chemical produced largely by certain bacteria and fungi that can inhibit or destroy the growth of other organisms, including pathogenic microorganisms.

Antimetabolite. A substance that inhibits the utilization of a metabolite necessary for growth.

Antiseptic. A chemical that inhibits or kills microbes. The definition also implies that the chemical is sufficiently nontoxic that it can be applied to skin and mucous membranes.

Coenzyme. Any heat-stable, nonprotein compound that forms an active portion of an enzyme system after combination with an enzyme precursor (apoenzyme). Many of the B vitamins are coenzymes.

Competitive inhibition. The inhibition of enzyme activity by competition between the inhibitor and the substrate for the active (catalytic) site on the enzyme.

Disinfectant. A chemical agent that rids an area of pathogenic microorganisms. In so doing, it kills vegetative forms of bacteria but ordinarily not endospore forms. The definition also implies that the chemical is sufficiently toxic that it should not be applied to body surfaces, only to material objects.

Essential metabolic pathway. A pathway of chemical transformations necessary for growth; if inhibited, the organism usually dies. The Krebs cycle and Embden-Meyerhof pathway are classic examples.

Essential metabolite. A chemical necessary for proper growth.

In vitro. Growth of microbes in test tubes.

In vivo. Growth of microbes in live plants and animals.

Pathogen. Any agent capable of causing disease, usually a microorganism.

Sepsis. The presence of pathogenic microorganisms or their toxins in tissue or blood; infection.

Objectives

1. To provide introductory information about the origin and use of antibiotics, antiseptics, and disinfectants.
2. To provide an opportunity to evaluate the inhibitory activity of antibiotics, antiseptics, and disinfectants on bacteria using a modified Kirby-Bauer test.

Reference

Nester et al. *Microbiology: A human perspective*, 6th ed. The McGraw-Hill Companies, 2009. Chapter 5, Section 5.4 and Chapter 21, Section 21.4.

Note: This exercise can be divided into two laboratory periods.

Materials

Cultures (per team of 2–4 students) of the following:

Bacteria (24 hour 37°C TS broth cultures)

Staphylococcus epidermidis (a Gram-positive coccus)

Escherichia coli (a Gram-negative rod)

Pseudomonas aeruginosa (nonfermenting Gram-negative rod)

Mycobacterium smegmatis (acid-fast rod)

Vials or dispensers of the following antibiotic discs:

Penicillin, 10 μg; streptomycin, 10 μg; tetracycline, 30 μg; chloramphenicol, 30 μg; sulfanilamide (or another sulfonamide), 300 μg; erythromycin, 15 μg

Beakers containing 10 ml aliquots of the following chemicals: 70% ethanol, 3% hydrogen peroxide, antiseptic mouthwash such as Listerine®, and disinfectant such as Lysol®

Mueller-Hinton agar, 6 plates

Sterile cotton swabs, 6

Sterile filter paper discs, ¼" or ½", 8

Small forceps, 1 per student

Ruler divided in mm

Procedure

First Session: Filter Paper Disc Technique for Antiseptics and Disinfectants

1. With a glass-marking pen, divide the underside of two plates of Mueller-Hinton agar into quadrants and label them 1–4.
2. Record codes for the four antiseptics and disinfectants on the bottom sides of the two agar plates, one code for each quadrant: 70% ethanol—E; 3% hydrogen peroxide—HP; Listerine®—L; and Lysol®—Ly.
3. Label the bottom of one petri dish *Staphylococcus epidermidis* and the bottom of the other dish *Escherichia coli*.
4. Suspend the *Staphylococcus epidermidis* culture, insert and moisten a sterile swab in the culture, remove excess, then streak the swab in three directions on the surface of the agar plate. Discard swab in the appropriate waste container.
5. Repeat step 4 with *Escherichia coli*.
6. Sterilize forceps by dipping them in 95% alcohol and then touch them to the flame of the Bunsen burner. Air cool.

7. Using forceps, remove one of the filter paper discs from the container and dip it into solution 1—70% ethanol.

8. Drain the disc thoroughly on a piece of clean absorbent toweling and place it in the center of quadrant 1 of the dish labeled *Staphylococcus epidermidis* (figure 14.2). Tap disc gently.

Figure 14.2 Filter paper disc technique for antiseptics and disinfectants.

(a) Antiseptic solution

Hold paper disc with forceps and dip into solution.

(b) Drain disc on toweling.

Place disc in center of quadrant. Repeat with three other antiseptics. Incubate dish at 37°C, 48 hours.

(c) Petri dish seeded with *S. aureus* or *E. coli*

(d) Measure the clear zone of inhibition surrounding each disc.

9. Repeat steps 5, 6, and 7 and place the disc in the center of quadrant 1 of the plate labeled *Escherichia coli*.

10. Repeat steps 5–8 for the remaining three compounds, using first 3% hydrogen peroxide, then the antiseptic mouthwash, and last Lysol®.

11. Invert the petri dishes and incubate at 37°C for 48 hours.

First Session: Filter Paper Disc Technique for Antibiotics

1. Divide the four broth cultures among team members so that each student sets up at least one susceptibility test.

2. With a glass-marking pen, divide the underside of four plates of Mueller-Hinton agar into six pie-shaped sections (figure 14.3a). Divide in four sections if using standard petri plates.

3. Record the codes of the six antibiotic discs on the bottom side of the four plates, one code for each section. See table 14.4 of the Laboratory Report for code designations.

4. Label the bottom of each plate with the name of the respective bacterium (see the Materials section for names).

5. Using aseptic technique, streak the first broth culture as shown in figure 14.3b. The remaining three cultures should be streaked on separate plates in a similar manner.

6. Heat sterilize forceps (see step 6 of the previous Procedure section), and remove an antibiotic disc from the container. Place gently, with identification side up, in the center of one of the pie-shaped sections of the agar plate (figure 14.3a). Tap gently to fix in position.

7. Continue placing the remaining five antibiotic discs in the same way.

   Note: Be sure to flame the forceps with alcohol after placing each disc because it is possible to contaminate stock vials with resistant organisms.

   Note: If a disc dispenser is used, follow the manufacturer's instructions.

8. Repeat steps 5–7 with the remaining three cultures.

9. Invert and incubate the plates at 37°C for 48 hours.

Figure 14.3 Antibiotic susceptibility test. *(a)* The underside of a Mueller-Hinton agar plate showing the marking of sections and the arrangement for placement of antibiotic discs on the agar surface. *(b)* Procedure for streaking an agar plate in three or more directions with a swab inoculum in order to achieve a uniform lawn of growth.

(a)

(b)

Second Session: Filter Paper Disc Technique for Antiseptics and Disinfectants

1. Turn over the *Staphylococcus epidermidis* plate, and with a ruler calibrated in mm, determine the diameter of the clear zone surrounding each disc. Repeat with the *Escherichia coli* plate.

   Note: It may be necessary to illuminate the plate in order to define the clear zone boundary.

2. Record your results in table 14.3 of the Laboratory Report.

Second Session: Filter Paper Disc Technique for Antibiotics

1. Observe plates using the same method described in step 1 of the second session for antiseptics and disinfectants. In addition, make note of any large colonies present in the clear zone of growth inhibition surrounding each antibiotic disc. They may be resistant mutants.

2. Record your findings in table 14.4 of the Laboratory Report.

3. Compare your results where possible with table 14.2 and indicate in table 14.4 the susceptibility of your test cultures (*when possible*) to the antibiotics as resistant (R), intermediate (I), or susceptible (S).

Note: Your answers may not agree exactly with those in table 14.2 since this is a modified Kirby-Bauer test.

EXERCISE

14

Laboratory Report: Antiseptics and Antibiotics

Results

Filter paper disc technique for antiseptics and disinfectants:

Table 14.3 Inhibitory Activity of Various Antiseptics and Disinfectants

Antiseptic or Disinfectant	Zone of Inhibition (mm)			
	Staphylococcus epidermidis	Optional Organisms	Escherichia coli	Optional Organisms
70% ethanol (E)				
3% hydrogen peroxide (HP)				
Listerine® (L)				
Lysol® (Ly)				
Others:				

1. What general conclusions can you make from this study? What differences, if any, did you observe on your plates between antiseptic and disinfectant preparations?

Filter paper disc technique for antibiotics:

Table 14.4   Antibiotic Susceptibility (Modified Kirby-Bauer Test)

Test Organism	Zone of Inhibition (mm)						Susceptibility[a]					
	Chl	Ery	Pen	Str	Sul	Tet	Chl	Ery	Pen	Str	Sul	Tet
Staphylococcus epidermidis												
Escherichia coli												
Pseudomonas aeruginosa												
Mycobacterium smegmatis												

[a]R = Resistant, I = Intermediate, and S = Susceptible.

Questions

1. What relationship did you find, if any, between the Gram-staining reaction of a microorganism and its susceptibility to antiseptics and disinfectants?

2. Did you observe any small colonies growing in the zone of inhibition around the antibiotic disc? Explain how this might occur.

3. How does an antibiotic differ from a disinfectant?

4. What else must be considered besides zone size when evaluating an antibiotic?

5. Under what circumstances would you use an antibiotic rather than a disinfectant?

6. Under what circumstances would you use a disinfectant rather than an antibiotic?

NOTES:

*I*NTRODUCTION to Microbial Genetics

In this section, three aspects of microbial genetics will be studied: selection of mutants, gene transfer, and gene regulation.

Selection of Mutants   Mutations are constantly occurring in all living things. The replication of DNA is amazingly error-free, but about once in every 100 million duplications of a gene a change is made. There are three possible outcomes:

1. There is no effect. Perhaps the altered base did not lead to structural change in a protein and the cell remained functional.
2. The mutation has affected a critical portion of an essential protein, resulting in the death of the cell.
3. The mutation has enabled the cell to grow faster or survive longer than the other nonmutated cells. This outcome rarely occurs.

Gene Transfer   Bacteria can transfer genetic material to other bacteria in three ways. These are conjugation, transduction, and transformation.

Conjugation occurs during cell-to-cell contact and is somewhat similar to sexual recombination seen in other organisms. The transferred DNA can be either chromosomal or a small, circular piece of DNA called a plasmid.

Transduction is the transfer of genes from one bacterial cell to another by a bacterial virus. These viruses, called bacteriophages or phages, package a bacterial gene along with the viral genes and transfer it to a new cell.

The third method of transferring genes is transformation, which is also called DNA-mediated transformation. (The word *transformation* is sometimes used to define the change of normal animal cells to malignant cells—a completely different system.) In bacterial transformation, isolated DNA is mixed with viable cells. It then enters the cells, which are able to express these new genes. Although it seems impossible for a large molecule such as DNA to enter through the cell wall and membrane of a living cell, this is indeed what happens.

Gene Regulation   Another aspect of genetics is the expression of genes. A cell must be economical with its energy and material and must not make enzymes or other products when they are not needed. On the other hand, a cell must be able to "turn on" genes when they are required in a particular environment. Gene regulation is examined in exercise 18.

NOTES:

EXERCISE 15

Selection of Bacterial Mutants Resistant to Antibiotics

Getting Started

All the bacterial cells in a pure culture are derived from a single cell. These cells, however, are not identical because all genes tend to mutate and form mutant organisms. The spontaneous **mutation rate** of genes varies between 1 in 10^4 to 1 in 10^{12} divisions, and even though they are quite a rare event, significant mutations are observed because bacterial populations are very large. In a bacterial suspension of 10^9 cells/ml, one could expect 10 mutations of a gene that mutated 1 in every 10^8 divisions.

Mutant bacteria usually do not grow as well as the **wild-type** normal cell because most changes are harmful, or at least not helpful. If, however, conditions change in the environment and favor a mutant cell, it will be able to outcompete and outgrow the cells that do not have the advantageous mutation. It is important to understand that the mutation is a random event that the cell cannot direct. No matter how useful a **mutation** may be in a certain situation, it just happens to the cell, randomly conferring an advantage or disadvantage to it.

In this exercise, you will select bacteria resistant to streptomycin. Streptomycin is an **antibiotic** that kills bacteria by acting on their ribosomes to prevent protein synthesis. (However, it does not stop protein synthesis in animals because eukaryotic ribosomes are larger than those of bacteria and therefore different.) **Sensitive** *E. coli* cells can become resistant to streptomycin with just one mutation.

In this exercise, you will select organisms resistant to streptomycin by adding a large population of sensitive bacteria to a bottle of TS broth containing streptomycin. Only organisms that already had a random mutation for streptomycin resistance will be able to survive and multiply (figure 15.1).

Figure 15.1 Selection of streptomycin-resistant *E. coli* cells.

A population of *E. coli* cells. Very few cells have mutated to streptomycin resistance.

Sensitive cell

Resistant cell

1 ml *E. coli*

1 ml *E. coli*

After incubation—sensitive cells + a few resistant cells

After incubation—resistant cells

TS broth

TS broth + streptomycin

Definitions

Antibiotic. A substance produced by one organism, usually a microorganism, that kills or inhibits other organisms.

Mutation. An inheritable change in the base sequences of DNA.

Mutation rate. The number of mutations per cell division.

Sensitive. An organism killed or inhibited by a particular antibiotic.

Wild type. The organism as it is isolated from nature.

Objectives

1. To understand the concept of selection and its relationship to mutation.
2. To understand that mutations are random events and that the cell cannot cause specific mutations to occur, no matter how advantageous they may be.
3. To count the number of streptomycin-resistant mutant bacteria that occur in an overnight culture of a sensitive strain.

Reference

Nester et al. *Microbiology: A human perspective,* 6th ed. The McGraw-Hill Companies, 2009. Chapter 8, Section 8.6.

Materials

Per team

First Session

    Flasks (or bottles) containing 50 ml TS broth, 2

    TS agar deeps, 2

    Sterile petri dishes, 2

    1-ml pipets, 2

    Overnight broth culture (~ 18 hours) of *Escherichia coli* K12 (about 10^9 cells/ml)

    Streptomycin solution at 30 mg/ml

Second Session

    TS agar deeps, 2

    Sterile petri dishes, 2

    1-ml pipets, 2

    Tubes of 0.5 ml sterile water, 2

    0.1-ml streptomycin

Procedure

First Session

1. Melt and place 2 TS agar deeps in a 50°C water bath.
2. Label one petri plate and one flask "with streptomycin." Label the other flask and plate "without streptomycin control" (figure 15.2).
3. Add 0.3 ml streptomycin to the flask labeled "streptomycin" and 0.1 ml to one of the melted, cooled agar deeps. Discard the pipet.
4. Immediately inoculate the TS agar deep with 1 ml of the bacterial culture, mix, and pour in the plate labeled "with streptomycin."
5. Add 1 ml of the bacteria to the tube of melted, cooled agar without streptomycin and pour into plate labeled "without streptomycin."
6. Add 1 ml of bacteria to each of the flasks.
7. Incubate the plates and flasks at 37°C. If using bottles, lay them on their side to increase aeration.

Second Session

1. Melt and cool two tubes of TS agar in a 50°C water bath.
2. Pour one tube of melted agar into a petri dish labeled "without streptomycin" and let harden.

Figure 15.2 Inoculating media with and without streptomycin with a culture of *E. coli* (session 1).

1 ml *E. coli*   1 ml *E. coli*   1 ml *E. coli*   1 ml *E. coli*

TS broth control   TS broth + streptomycin   TS agar control   TS agar + streptomycin

3. Add 0.1 ml streptomycin to the other tube of melted agar, pour into a petri dish labeled "with streptomycin," and let harden.
4. Examine the bottles and plates inoculated during the first session. Note whether there is growth (turbidity) or not in both of the bottles. Count the number of colonies growing in the pour plates. How many streptomycin-resistant mutants/ml were present in the original inoculum? Compare it to the growth of organisms in the control plate without streptomycin. If there are more than 300 colonies or the plate is covered by confluent growth, record as TNTC or "too numerous to count." Record results.
5. Test the bacteria growing in the bottles and on the plates for sensitivity or resistance to streptomycin in the following way: Divide both agar plates in four sections as diagrammed in figure 15.3, then take a loopful of broth from the bottle without streptomycin and inoculate

a sector of each agar plate. Do the same with the broth culture containing streptomycin.
6. Dig an isolated colony out of the agar plate containing the streptomycin and suspend it in a tube of sterile water. Use a loopful to inoculate the third sector of each plate. Also suspend some organisms from the control plate in sterile water (there will not be any isolated colonies) and inoculate the fourth sector. Incubate the plates at 37°C.
7. Predict which bacteria will be sensitive to streptomycin and which will be resistant.

Third Session

1. Observe the growth on each sector of the plates and record results. Were they as you predicted?
2. Occasionally, mutants will not only be resistant to streptomycin, but also will require it. If you have one of these unusual mutants, be sure to show it to the instructor.

Figure 15.3 Testing previously incubated cultures for streptomycin sensitivity (session 2).

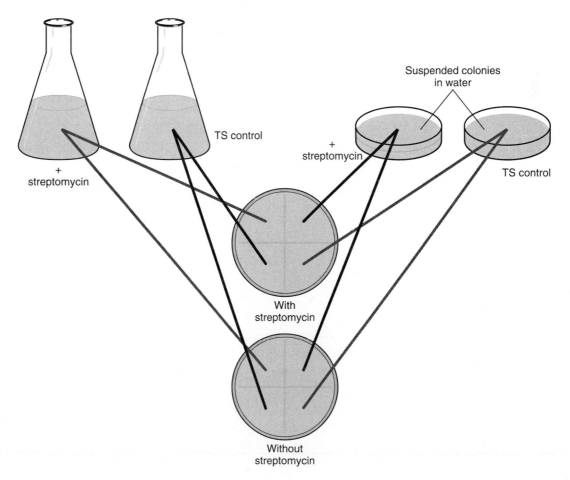

NOTES:

EXERCISE

15

Laboratory Report:
Selection of Bacterial Mutants Resistant to Antibiotics

Results (Second Session)

Source	Growth / No Growth
TS broth (control)	
TS broth plus streptomycin	

Source	Number of Colonies
TS agar plate (control)	
TS agar plate plus streptomycin	

Results (Third Session)

Source	Growth on TS ager Plate	Growth on TS ager Plate + Strp
TS broth (control)		
TS broth plus streptomycin		
TS agar (control)		
TS agar plus streptomycin		

How many organisms/ml were streptomycin resistant in the original overnight culture of sensitive *E. coli?*

Questions

1. Two bottles of TS broth (with and without streptomycin) were inoculated in session 1 with 1 ml of an overnight culture of *Escherichia coli*. After incubation, why was one streptomycin population sensitive and the other streptomycin population resistant?

2. How were you able to estimate the number of streptomycin-resistant organisms already present in the overnight culture of *Escherichia coli* growing in the TS broth?

3. Why should antibiotics only be used if they are necessary?

4. Which is correct?
   a. An organism becomes resistant after it is exposed to an antibiotic.

   b. An antibiotic selects organisms that are already resistant.

EXERCISE

16

Transformation: A Form of Genetic Recombination

Getting Started

In this exercise, transformation is used to transfer the genes of one bacterium to another. It gives you a chance to see the results of what seems to be an impossible process—a huge DNA molecule entering an intact cell and permanently changing its genetic makeup.

Basically, the process involves mixing DNA from one strain of lysed (disrupted) cells with another strain of living cells. The DNA then enters the viable cells and is incorporated into the bacterial chromosome. The new DNA is expressed, and the genetic capability of the cell may be changed.

In order to determine whether the bacteria are indeed taking up additional DNA, the two sets of organisms (DNA donors and DNA recipients) must differ in some way. One strain usually has a "marker," such as resistance to an antibiotic or the inability to synthesize an amino acid or vitamin. In this exercise, a gene responsible for conferring resistance to the antibiotic streptomycin is transferred to cells that are sensitive to it (figure 16.1).

The organism used in this exercise is *Acinetobacter* (ā sin NEET o bacter), a short, Gram-negative rod found in soil and water. The prefix "a" means without, and "cine" means movement, as in cinema; thus *Acinetobacter* is nonmotile. This organism is always **competent,** which means it can always take up **naked DNA.** Most bacteria are not competent unless they are in a particular part of the growth curve or in a special physiological condition. In any event, for transformation to occur, the DNA must not be degraded (chemically broken down). If an enzyme such as **DNase** is present, it cuts the DNA in small pieces, preventing transformation.

Definitions

Competent. Cells that are able to take up naked DNA.

Figure 16.1 Transformation of cells with a gene conferring streptomycin resistance.

Gene conferring streptomycin resistance

Bacterial chromosome

Cells resistant to streptomycin

Cells lysed, releasing DNA

DNA mixed with sensitive cells

Previously sensitive cells are now streptomycin resistant. They have been transformed.

DNase. An enzyme that cuts DNA, making it useless for transformation.

Naked DNA. DNA released from lysed, or disrupted, cells and no longer protected by an intact cell.

Objectives

1. To understand the process of transformation, and to observe it in the laboratory.
2. To understand the use of genetic markers.
3. To understand the importance of controls in an experiment.

Reference

Nester et al. *Microbiology: A human perspective,* 6th ed. The McGraw-Hill Companies, 2009. Chapter 8, Section 8.7.

Safety Precaution: *Acinetobacter* can cause pneumonia in immunologically compromised individuals.

Materials

Per team

    TSY (trypticase soy yeast extract) agar plate, 1

    TSY agar plate with streptomycin (second session), 1

    Broth culture of *Acinetobacter* Str^R (resistant to streptomycin), 1

    Broth culture of *Acinetobacter* Str^S (sensitive to streptomycin), 1

    Tube with 0.1 ml detergent SDS (sodium dodecyl sulfate) in 10× saline citrate, 1

    Solution of DNase

    1-ml pipet, 1

Class equipment

    60°C water bath with test tube rack

Procedure

First Session

1. Transfer 1.0 ml of Str^R *Acinetobacter* broth culture into the tube of SDS. Label the tube and incubate it in the 60°C water bath for 30 minutes. The detergent (SDS) will lyse the cells, releasing DNA and other cell contents. Any cells that are not lysed will be killed by the 30-minute exposure to 60°C water. Label the tube "DNA."

2. Divide the bottom of the TSY agar plate into five sectors using a marking pen. Label the sections "DNA," "Str^S (streptomycin sensitive)," "Str^R (streptomycin resistant)," "Str^S + DNA," and "Str^S + DNase + DNA" (figure 16.2).

3. Inoculate the plate as indicated by adding a loop of the broth culture, DNA, or DNase in an area about the size of a dime to each sector.
   a. DNA. The lysed mixture of Str^R cells is the source of DNA. It also contains RNA, proteins, and all the other cell components of the lysed cells that do not interfere with the transformation.
   b. Str^S cells. Inoculate a loopful of the Str^S culture.
   c. Str^R. Inoculate a loopful of the Str^R culture.

Figure 16.2 Five labeled sectors of a TSY agar plate.

d. Str^S cells + DNA. Inoculate a loopful of Str^S cells and add a loopful of the DNA (lysed Str^R cells) in the same area. THIS IS THE ACTUAL TRANSFORMATION. Str^R cells will grow here if transformed by the DNA.

e. Inoculate a loopful of Str^S cells as above, and in the same area add a loopful of DNase, then add a loopful of DNA. It is important to add these in the correct order or transformation will occur before the DNase can be added. The DNase should degrade the DNA and prevent transformation.

4. Incubate the plates at room temperature for several days or at 37°C for 48 hours.

Second Session

1. Observe the plate you prepared in the first session. There should be growth in all sectors of the plate except the DNA sector (*a*). If

the DNA control sector shows growth, it indicates that your crude DNA preparation was not sterile but contained viable cells. If this has happened, discard your plates and borrow another student's plate after he or she is finished with it; there should be sufficient material for more than one team. Why is it so important that the DNA preparation is sterile?

2. Be sure you understand the purpose of each control. Which sector demonstrates that
   a. the Str^S cells were viable and can grow on the agar plate?
   b. the Str^R cells were viable?
   c. the lysed mixture of Str^R cells contained no viable organisms?
   d. DNA is the component of the lysed cells responsible for transformation?

3. Divide the bottom of a TSY + streptomycin plate into four sectors and label them "Str^S," "Str^R," "Str^S + DNA," and "Str^S + DNase + DNA."

4. Streak a loopful of cells from the first plate to the corresponding sectors on the TSY + streptomycin plate. Lightly spread them in an area about the size of a dime. Cells growing on this plate must be streptomycin resistant.

5. Incubate at room temperature for several days or at 37°C for 48 hours or until cells have grown.

Third Session

1. Observe the TSY + streptomycin agar plate inoculated last session and record results. Did you transform the cells sensitive to streptomycin to cells that were resistant and could now grow on streptomycin?

2. Did the results from the controls confirm that naked DNA can transfer resistance to streptomycin?

NOTES:

EXERCISE

16

Laboratory Report: Transformation:
A Form of Genetic Recombination

Results

Indicate growth (+) or no growth (−) in each sector.

Results of first session

TSY agar

Results of second session

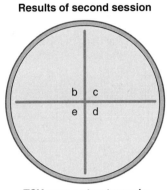

TSY agar + streptomycin

	Yes/No	Plate and Sector that Demonstrates Observation
Were *Acinetobacter* Str^S cells sensitive to streptomycin?		
Were *Acinetobacter* Str^R cells resistant to streptomycin?		
Was the DNA (cell lysate) free of viable cells?		
Did transformation take place?		
Did the DNase prevent transformation?		

Questions

1. What two components were mixed together to show transformation?

2. What is the action of DNase?

3. What control showed that transformation—not conjugation or transduction—was responsible for the results?

4. If the Str^S cells had grown on the TSY + streptomycin agar plate, would you have been able to determine if transformation had taken place? Explain.

5. If you had used a DNA lysate containing viable cells, would it have been possible to determine whether transformation had taken place? Explain.

6. How does transformation differ from conjugation and transduction?

EXERCISE

17

Bacterial Conjugation

Getting Started

Conjugation is one of the three mechanisms responsible for genetic transfer in bacteria (see Introduction). In conjugation, a cell of mating type F$^+$ attaches by a pilus to a cell of mating type F$^-$, and DNA is transferred from the F$^+$ cell to the F$^-$.

The donor cell is designated **F$^+$** because it contains a fertility factor known as an **F factor** and the recipient is **F$^-$** because it lacks the F factor. Some of the genes that make up the F factor code for the sex pilus, an appendage on the surface of the cell.

The F factor is usually located on a **plasmid,** which is a small, circular piece of DNA found in the cytoplasm of the cell. The genes found on plasmids are usually helpful to the cell in certain situations, but are not essential for the cell's normal metabolism. For example, the genes conferring a certain type of resistance to antibiotics are frequently carried on plasmids.

During conjugation, the pilus of the F$^+$ cell attaches to the F$^-$ cell and seems to be involved in bringing the cells into close contact. One strand of the plasmid enters the F$^-$ cell, the plasmid DNA replicates, and the F$^-$ cell becomes an F$^+$ cell. The cell can then express all the other genes (such as antibiotic resistance) contained on the plasmid. Conjugation is therefore a very important mechanism in the spread of antibiotic-resistant genes (figure 17.1*a*).

F factors can be responsible for the transfer of more than just the genes on a plasmid. The F factor can, on rare occasions, integrate into the bacterial chromosome, which can then be transferred during conjugation (see figure 17.1*b*). These strains are called **Hfr** strains for *h*igh *f*requency of *r*ecombination.

In this exercise, a donor Hfr strain containing the F factor in the chromosome will be used to transfer chromosomal genes to an F$^-$ strain of *E. coli.* Each of the strains must have a genetic "marker," such as antibiotic resistance or the inability to

synthesize an essential cellular metabolite (*e.g.,* an amino acid or a vitamin). In this case, the donor strain is unable to synthesize the amino acid methionine and the recipient cannot synthesize the amino acid threonine. Neither strain can grow on a mineral medium because it does not contain the amino acids that the strains require. (Mutants requiring a growth factor are termed **auxotrophs.**) If, however, the normal genes are transferred to the mutant recipient during conjugation, these cells then will be able to synthesize all the required amino acids. Recombinant bacteria will be able to grow on the mineral medium but the auxotrophic parent strains will not.

During conjugation, the chromosome is transferred in a linear manner. The longer the cells stay in contact, the more chromosome that is transferred, but the cells usually break apart before the whole chromosome is transferred. It is important not to shake the culture during conjugation so that cells stay in contact as long as possible.

Note: Auxotrophic organisms can sometimes regain the ability to synthesize an amino acid by having a second mutation, or **back mutation,** that restores the activity of the gene. This is a rare event (around an average of 10^{-8} or 1 in 100,000,000 cell divisions).

Definitions

Auxotroph. An organism that cannot synthesize all its needed growth factors. The strain labeled methionine in the conjugation exercise is an auxotroph requiring methionine. It cannot grow on a mineral medium, but only on TS agar or another medium that contains methionine.

Back mutation (also called *reversion*). A second mutation restoring the activity of a mutated gene.

Figure 17.1 *(a)* Conjugation: transfer of the F⁺ plasmid. *(b)* Conjugation: transfer of chromosomal DNA.

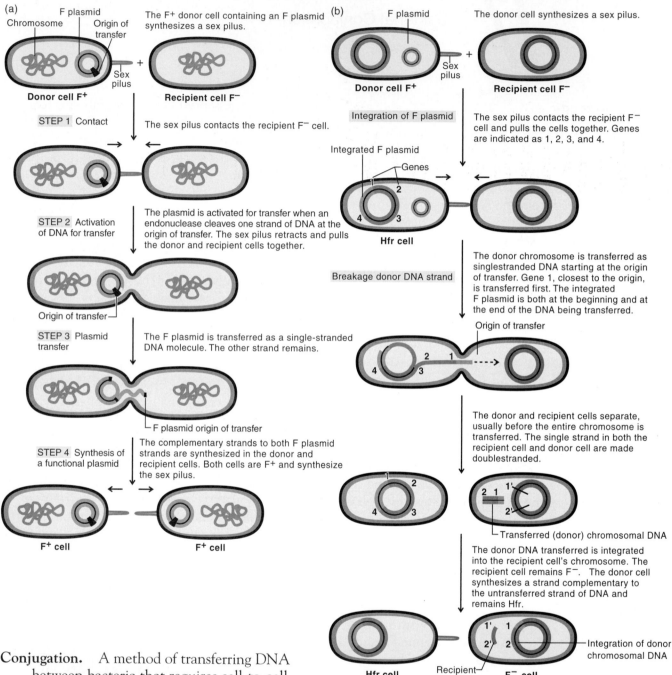

Conjugation. A method of transferring DNA between bacteria that requires cell-to-cell contact.

F factor. (F plasmid) Genes giving the cell the ability to transfer DNA via conjugation.

F⁻. Cells lacking the F factor; called the *recipient bacteria.*

F⁺. Cells containing the F factor; called the *donor bacteria.*

Hfr. (*High frequency of recombination*) F factor is incorporated in the chromosome of the bacteria, and consequently portions of the chromosome can be transferred to F⁻ bacteria.

Plasmid. A small circle of DNA found in some cells in addition to the chromosomal DNA.

Objectives

1. To increase knowledge of concepts and techniques used in the study of genetics.
2. To acquaint you with a laboratory method for demonstrating conjugation using auxotrophic organisms.
3. To demonstrate the importance of conjugation in transferring genes among bacteria.

Reference

Nester et al. *Microbiology: A human perspective,* 6th ed. The McGraw-Hill Companies, 2009. Chapter 8, Section 8.8.

Materials

Per team

TS broth culture in log phase of *E. coli* A F$^+$ methionine-(Hfr donor)

TS broth culture in log phase of *E. coli* B F$^-$ threonine-(recipient)

TS agar plates, 1

Sterile test tube, 1

Sterile 9-ml water blanks, 5

Agar plates of mineral salts + 0.5% glucose (mineral medium, or MM), 4

Sterile 1.0-ml pipets, 6

Sterile bent glass rod (sometimes called dally rods or hockey sticks), 1

Procedure

Steps are outlined in figure 17.2.

First Session

1. Label all tubes and plates as indicated in figure 17.2.
2. Transfer 1 ml of culture A into the water blank labeled A 10^{-1}. With the same pipet, transfer a second 1 ml of culture A into the tube labeled A + B.
3. Transfer 1 ml of culture B into the water blank labeled B 10^{-1}. With the same pipet, transfer a second 1 ml of culture B into the tube labeled A + B.
4. Gently mix the A + B tube and incubate it without shaking for 30 minutes at room temperature. This is the actual mating.
5. While the A + B mixture is incubating, inoculate half of the MM plate (labeled A) and half of the TS agar plate (labeled A) with loopfuls of culture A. Repeat with culture B. These are controls to verify that neither culture A nor B can grow on MM agar, but can grow on TS agar. These cultures were diluted 1/10 so that you do not carry over any of the rich TS broth medium when testing the organisms on the mineral medium.
6. After 30 minutes' incubation, transfer 1 ml of the A + B mixture into the water blank labeled A + B 10^{-1}. Mix thoroughly and transfer 1 ml with a second sterile pipet to the water blank labeled A + B 10^{-2}. Again, mix thoroughly and transfer 1 ml with a third sterile pipet to the water blank labeled A + B 10^{-3}. Mix thoroughly.
7. With a sterile 1.0-ml pipet remove 0.1 ml from the A + B 10^{-3} dilution and add it to the surface of the MM agar plate labeled A + B 10^{-3}. Immediately spread the drop completely over the surface of the plate with the sterile bent glass rod.
8. Using the same pipet and bent glass rod, repeat the procedure to inoculate the MM agar plates labeled A + B 10^{-2} and A + B 10^{-1} from their respective dilutions. Remember, you can use the same pipet when going from low concentrations to higher concentrations.
9. Invert the agar plates and incubate at 37°C for 2–3 days.

Second Session

1. Examine the control MM agar and TS agar plates and record results. Are both cultures A and B viable auxotrophs?
2. Count the number of colonies on the plates that have countable numbers. Record the results.

Figure 17.2 Procedure for conjugation exercise.

Conjugation

Controls

Use the same tubes of *E. coli* A and *E. coli* B shown above.

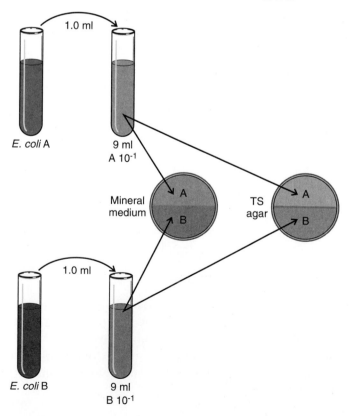

EXERCISE

17

Laboratory Report: Bacterial Conjugation

Results

Indicate where growth occurred.

	Culture A	Culture B
TS agar		
MM agar		
Culture an auxotroph?		

Plate Counts	A + B 10^{-1}	A + B 10^{-2}	A + B 10^{-3}
Number of colonies			

Summary:

	Genotype	Phenotype
Donor cells		
Recipient cells		
Recombinant cells		

1. How many recombinant organisms resulted from the mating? Show calculations.

2. If there were roughly 2×10^8 organisms in the A + B mixture, what percent of the original mixture resulted in recombinants?

Note: Since only the recipient cells can receive DNA and change genotype, the percent of recombinants should be based on the number of recipient cells, or 1×10^8 cells/ml.

3. If the back mutation rate for threonine is about 1 in 10^8 cells, about how many back mutants would you expect to be present per ml in the A + B mixture?

Questions

1. Why can the same pipet and bent glass rod be used to inoculate plates when starting with the most dilute mixture?

2. Why is an auxotrophic organism not able to grow on MM agar?

3. If either A or B could grow on the MM agar, how would the results change?

4. Using the answers you calculated in part 2 in the Results section, how did the number of recombinants resulting from conjugation compare with the number you expect from back mutation?

5. Compare the effects of mutation and conjugation to produce organisms with new genetic capability.

NOTES:

EXERCISE 18

Gene Regulation: Induction and Catabolite Repression

Getting Started

A bacterial cell has all the genetic information to produce and operate a new cell. This includes the enzymes necessary for obtaining energy and synthesizing the necessary cellular components. A cell must work as efficiently as possible and carefully utilize the available nutrients without investing energy in enzymes not needed. Some enzymes are utilized in the basic energy pathways of the cell and are continually synthesized. These are called **constitutive enzymes.**

Others, termed **inducible enzymes,** are only needed when their specific **substrate** is available. For example, *E. coli* can break down the sugar lactose with an enzyme called β-galactosidase. If lactose were not present in the environment, it would be a waste of energy and of intermediate compounds to synthesize this enzyme. Therefore, these kinds of enzymes are called inducible enzymes because the presence of the substrate induces their synthesis.

How does the cell control inducible enzymes? In bacteria, this takes place on the level of **transcription.** Inducible enzymes can be found in an **operon,** which has a promoter and operator followed by the genes involved in the enzymatic activity (figure 18.1). The repressor binds to the operator gene, blocking transcription if the substrate for the enzyme is not present. When the substrate is present, it binds to the repressor, allowing the RNA polymerase to transcribe the genes for β-galactosidase. The genes involved with lactose utilization are in a cluster termed the lac operon.

Another method of control utilized by the cell is **catabolite repression,** which occurs when a cell has a choice of two sources of energy, one of which is more easily utilized than the other. An example is the presence of both glucose and starch. Glucose can immediately enter the glycolytic pathway, while starch must first be cleaved with amylase. Amylases and other enzymes cost the cell energy and materials to produce, so therefore it is much more economical for the cell to utilize glucose if it is present. When glucose is present along with starch, the glucose represses the synthesis of amylase even though the enzyme would normally be induced in the presence of the starch.

The first procedure of this exercise will investigate induction of β-galactosidase, a particularly important enzyme system used in recombinant DNA techniques as a measure of gene expression. The enzyme β-galactosidase cleaves the disaccharide lactose into glucose and galactose. It would be inefficient for the cell to synthesize this enzyme if lactose were not present, so therefore only the presence of lactose induces the synthesis of β-galactosidase.

However, when actually testing this reaction, it is difficult to detect the presence of β-galactosidase, a colorless protein. This problem has been solved by adding **ONPG** (o-nitrophenyl-β-D-galactoside). Most enzymes are specific for just one substrate or reaction, but β-galactosidase can also cleave ONPG, a synthetic compound, forming a yellow dye. Therefore a positive test for β-galactosidase is a yellow color in the presence of ONPG.

The second procedure of this exercise will test for the catabolite repression of amylase. Amylase is an exoenzyme; it is excreted outside the cell because the large starch molecule may be too large to pass through the cell membranes. The presence of starch is detected by adding iodine, which turns the starch purple.

Definitions

Catabolite repression. The cell's mechanism for preventing the synthesis of enzymes needed to break down an energy source if a simpler energy source, such as glucose, is available.

Constitutive enzymes. Enzymes continually produced by the cell.

Inducible enzymes. Enzymes produced only when substrate is present.

Figure 18.1 Inducible enzymes. Regulation of enzyme synthesis in a degradative enzyme system. Note that the same genetic elements take part as in the regulation of a biosynthetic pathway.

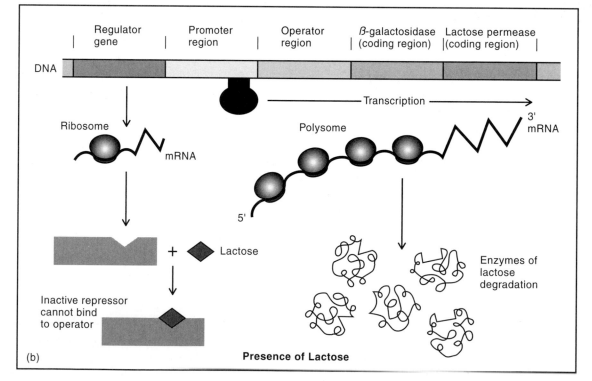

ONPG (o-nitrophenyl-β-D-galactoside). A compound used for detecting the presence of β-galactosidase.

Operon. A series of genes controlled by one operator (gene).

Substrate. The molecule reacting with an enzyme.

Transcription. Transfer of the genetic information from DNA to messenger RNA.

Objectives

1. To understand the concepts of induction.
2. To understand the lac operon and the use of ONPG.
3. To understand the concept of catabolite repression and how it is tested.

Reference

Nester et al. *Microbiology: A human perspective*, 6th ed. The McGraw-Hill Companies, 2009. Chapter 7, Section 7.5.

Induction

Materials

Per pair of students

Mineral salts + 0.2% glucose broth
5 ml/tube

Mineral salts + 0.2% lactose broth
5 ml/tube

Mineral salts + 0.2% glycerol broth
5 ml/tube

3 ml of ONPG

Cultures of the following:

Overnight TS broth cultures of *Escherichia coli* (cultures of *Enterobacter* and *Klebsiella* may also be tested)

Procedure

First Session

1. Inoculate each tube with a drop or loopful of *E. coli*.
2. Incubate at 37°C until the next laboratory period or at least 48 hours.

Second Session

1. Examine the tubes for growth—all the tubes should be turbid.
2. Add 1 ml of ONPG to each tube. The indicator compound ONPG is cleaved by β-galactosidase into a yellow product.
3. Incubate at room temperature for 30 minutes.

4. Examine the tubes to determine if the broth has turned yellow—an indication of the presence of the induced enzyme β-galactosidase.
5. Record results.

Catabolite Repression

Materials

Cultures of the following:

Bacillus growing on TS agar slants

Nutrient + starch agar plate

Nutrient + starch + glucose agar plate

Gram's iodine

Procedure

First Session

1. Label each plate.
2. Inoculate the middle of each plate with the *Bacillus* in an area about a few mm square.
3. Incubate at 30°C for a day or two. Try not to let the *Bacillus* grow over more than a third or half the plate. You may have to refrigerate the plates if the colony becomes too large.

Second Session

1. Flood the agar plates with Gram's iodine. The starch will turn purple. If the starch has been broken down with amylases, a clear zone will appear around the colony (figure 18.2).
2. Record results.

Figure 18.2 Appearance of plates after flooding with Gram's iodine.

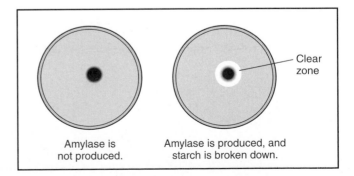

Amylase is not produced.

Amylase is produced, and starch is broken down.

Clear zone

NOTES:

EXERCISE

18

Laboratory Report: Gene Regulation:
Induction and Catabolite Repression

Results

Results of induction

	Glucose	Lactose	Glycerol
Color			
β-galactosidase present?			

Results of catabolite repression

	Starch	Starch + Glucose
Zone of clearing present?		
Amylase present?		

Questions

1. Which substrate induced β-galactosidase?

2. What reaction produced the yellow color?

3. What results would you expect if β-galactosidase were a constitutive enzyme?

4. In the catabolite repression exercise, did the *Bacillus* have the capacity to synthesize amylase (amylase +)? How did you determine your answer?

5. Did you observe catabolite repression when glucose was added to the starch? How did you determine your answer?

6. What results would you expect if amylase were a constitutive enzyme?

Introduction to the Other Microbial World

The phrase, the "other microbial world" refers to organisms other than bacteria, the major organisms of study in other parts of the manual. All of the organisms included here, with the exception of viruses, are eukaryotic organisms, many of which are of medical importance. Included are members of the nonfilamentous fungi (primarily yeasts), the filamentous fungi (molds), and intestinal animal parasites of medical importance, the protozoa and helminths. Viruses that infect both prokaryotes (bacteria) and eukaryotes (animal and plant cells) are introduced. Only bacterial viruses will be made available for laboratory study.

Mycology, or the study of fungi, is the subject of exercise 19. Included are a wide variety of fungal forms, ranging from unicellular microscopic cells, such as yeast, to filamentous multicellular structures. Many filamentous fungi, such as mushrooms, puffballs, toadstools, bracket fungi, and molds, are visible with the naked eye.

Most yeasts and filamentous fungi found in nature are nonpathogenic. In fact, many contribute to our well-being: for example, the yeast *Saccharomyces* is important for the manufacture of bread, beer, and wine; and the filamentous fungus *Penicillium chrysogenum* produces the antibiotic penicillin. Some are associated with spoilage (moldy jam and bread, mildew of clothing, etching of glass lenses in warm, humid environments) and, more recently, with production of food toxins known as aflatoxins by certain species of the genus *Aspergillus*.

Fungi are eukaryotic organisms that differ from algae in that they lack chlorophyll. Fungi differ from bacteria in that the cells are larger and contain membrane-bound organelles. In addition, bacteria are prokaryotes.

The fact that fungi are eukaryotic and bacteria are prokaryotic, in turn, affects antibiotic therapy because antibiotics effective against prokaryotes are usually ineffective against eukaryotes. If an antibiotic is effective against a fungus it may also, depending on its mode of action, damage the human host, because humans are also eukaryotic. Consequently,

antibiotic control of fungal infections is usually more difficult than control of bacterial infections. Fortunately, many fungal infections are opportunistic[1] infections so that healthy individuals rarely acquire them other than, perhaps, cutaneous fungal infections, such as athlete's foot. Nevertheless, there are a number of important mycotic diseases with which you should familiarize yourself.

Fungi can be cultivated in the laboratory in the same manner as bacteria. Physiologically, all fungi are heterotrophs (they require an organic source of carbon, such as glucose). Most are aerobic, although some are facultative (able to carry out both aerobic and anaerobic metabolism), and a few are obligate anaerobes (found in the rumens of cows). Most fungi grow best at temperatures of 20°–30°C, although some grow well at temperatures as high as 45°–50°C (such as *Aspergillus fumigatus*, an opportunistic filamentous fungus known to cause pulmonary aspergillosis).

Parasitic diseases constitute a major worldwide public health problem, both in developed and developing countries. In developing countries, parasitic diseases are prevalent due to poverty, malnutrition, lack of sanitation, and lack of education. A simple family survey for intestinal parasites in a small Central American village revealed that every family member harbored at least three types of parasites. Effects of human parasitic disease range from minimal, with some nutritional loss but only minor discomfort (common in pinworm and *Ascaris* infections), to debilitating and life-threatening disorders, such as malaria and schistosomiasis.

Industrialized countries also suffer from parasitic infections. Fecal contamination of drinking water by wild animal carriers, such as beavers, has in recent years caused major outbreaks of **giardiasis** (an intestinal disease) in various parts of the United

---

[1]Opportunistic infections are associated with debilitating diseases (such as cancer) and use of cytotoxic drugs, broad-spectrum antibiotics, and radiation therapy, all of which can suppress the normal immune response.

States. The causative agent, the protozoan *Giardia lamblia* (see figure 32.3), produces **cysts** that are quite small, thereby enabling them to sometimes pass through faulty water supply filters. The cysts often resist chlorination. The protozoan flagellate *Trichomonas vaginalis* is a common cause of vaginitis in women and is often sexually transmitted. Pinworm infestations are a problem among elementary schoolchildren. About 20% of infections in domestic animals are caused by protozoan and helminthic (worm) agents.

Because of their worldwide public health importance and their natural history differences from bacteria and fungi, we believe protozoa and helminths merit their own laboratory session. Exercise 20 provides insights on the diseases they cause and the techniques used for their diagnosis and identification.

Virology, the study of viruses (the word for poison in Greek), has early roots, although somewhat mysterious. The mystery stems, in part, from the fact that viruses could not be seen, even with a light microscope, and yet when sap from the leaf of an infected tobacco plant was passed through a filter that removed bacteria and fungi, the clear filtrate retained its infectious properties. It was not until the mid-1930s, with the advent and aid of the electron microscope, that viruses were first observed. In 1935, Wendell Stanley succeeded in crystallizing tobacco mosaic virus (TMV), enabling him to observe that it was structurally different from living cells (figure I.7.1).

Bacterial viruses, known as **bacteriophage,** were first described by Twort (1915) and later by d'Herelle (1917). d'Herelle observed their filterable nature and their ability to form **plaques** on an agar plate seeded with a lawn of the host bacterium (see figure 21.2). Both Twort and d'Herelle worked with coliform bacteria isolated from the intestinal tract.

The discovery of plaque formation on an agar medium was a godsend for future virology research in that it provided a fast, easy way to recognize, identify, and quantify bacteriophage infections. Viruses that attack mammalian cells also form structures analogous to plaques when cultivated on growth media able to support mammalian cell growth. Rather than plaques, the structures formed are described as **cytopathic effects (CPE)** (figure I.7.2). The CPE observed are dependent

Figure I.7.1   Tobacco mosaic virus. Electron micrograph (X approximately 70,000). Compare length and width with a rod-shaped bacterium. © Omikron/Photo Researchers, Inc.

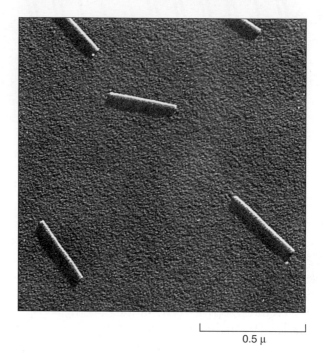

0.5 μ

Figure I.7.2   Mammalian virus plaques showing different cytopathic effects (CPE). Photographed 5 days after infection of the growth medium (a single layer of monkey kidney cells) with the various mammalian viruses. When in monolayers, the viruses are able to form plaques (a form of CPE), which can be detected macroscopically.

From *Diagnostic Procedures for Viral. Rickettsial and Chlamydial Infections,* 5th Edition. Copyright © 1979 by the American Public Health Association. Reprinted with permission.

upon the nature of both the host cell and the invading virus. The same can be said about the nature of bacterial plaque formation.

The advantage of using a bacteriophage to demonstrate virus growth is the ease of culturing the host cells. Mammalian cells require complex growth media, including blood serum, as well as prolonged incubation with the virus before CPE are visible. Therefore, we will work with an *E. coli* bacteriophage that can be obtained from a pure culture collection (see exercise 21).

Having made sundry efforts, from time to time, to discover, if 'twere possible, the cause of the hotness or power whereby pepper affects the tongue (more especially because we find that even though pepper hath lain a whole year in vinegar, it yet retaineth its pungency); I did now place anew about ⅓ ounce of whole pepper in water, and set it in my closet, with no other design than to soften the pepper, that I could the better study it. This pepper having lain about three weeks in the water, and on two several occasions snow-water having been added thereto, because the water had evaporated away; by chance observing this water on the 24th of April, 1676, I saw therein, with great wonder, incredibly many very little animalcules, of divers sorts; and among others, some that were 3 or 4 times as long as broad, though their whole thickness was not, in my judgment, much thicker than one of the hairs wherewith the body of a louse is beset. . . . The second sort of animalcules consisted of a perfect oval. They had no less nimble a motion than the animalcules first described, but they were in much greater numbers. And there was also a third sort,

which exceeded both the former sorts in number. These were little animals with tails, like those that I've said were in rainwater.

The fourth sort of little animals, which drifted among the three sorts aforesaid, were incredibly small; nay, so small, in my sight, that I judged that even if 100 of these very wee animals lay stretched out one against another, they could not reach to the length of a grain of coarse sand; and if this be true, then ten thousand of these living creatures could scarce equal the bulk of a coarse sand-grain.

I discovered yet a fifth sort, which had about the thickness of the last-said animalcules, but which were near twice as long.

DOBELL, *Antony van Leeuwenhoek and His Little Animals*

Definitions

Bacteriophage. A virus which infects bacteria. Frequently termed phage.

Cyst. A resting structure found in some protozoa.

Cytopathic effects (CPE). Areas of clearing in tissue culture cells caused by the infecting animal virus.

Giardiasis. Intestinal disease caused by *Giardia lamblia* found in contaminated drinking water. Sometimes called beaver fever because it can be spread by wild animals.

Mycology. The study of fungi.

NOTES:

EXERCISE

19

Microscopic Identification of Fungi

Getting Started

When you hear the word "fungus," what does it suggest? If unacquainted with fungus nomenclature, it may make you think of mushrooms and **toadstools** or moldy fruit and yeast. All of these are examples of fungi. Fungi such as mushrooms and toadstools are macroscopic fungi, which can usually be identified without the aid of a microscope, whereas yeast and molds require microscopy for identification. However, since fungi are considerably larger than bacteria they may be easier to identify.

Molds, yeasts, and perhaps another group, the **lichens,** are all members of the true fungi (Eumycota). The lichens are placed with the fungi for convenience because they represent dual thallus

plants composed of an alga and a fungus. The two fungus subgroups are the **nonfilamentous fungi** (exemplified by the yeasts), which are unicellular and the **filamentous fungi** (exemplified by the molds), which are multicellular and have true filaments **(hyphae).** The hyphae are either nonseptate **(coenocytic)** or septate (figure 19.1). The nonseptate filaments are multinucleate, whereas the septate filaments contain either one or more nuclei per unit. This structural difference is important taxonomically in that one of the four classes of fungi, the Zygomycetes (table 19.1), is distinguished from the other classes by its lack of septate hyphae. Also in contrast to the other three classes, it contains only a few human pathogens but numerous plant pathogens. Some authors divide the Zygomycetes into two classes, the Zygomycetes and the Oomycetes. The Zygomycetes are terrestrial fungi, and the Oomycetes are aquatic fungi containing the preponderance of plant pathogens. Fungus classification, although still somewhat in a state of flux, continues to improve with time.

Why all this interest in fungi? As with other forms of life there are both the good and the bad fungi. Most are **saprophytes,** meaning that in the chain of nature they decompose dead matter into a form that can be used to support all sorts of living matter. There are also the fungi that cause disease in plants or animals. For example, studies show that fungi unable to synthesize certain of their own nutrients invade a plant for these nutrients and thereby destroy it. For this purpose, they have spearlike hyphae that are adapted for invasive growth. Strangely enough the same type of hyphae are also involved in the formation of various multicellular organs. The multicellular organs in turn can regenerate hyphae (see Moore, 1998). Today there is mounting interest worldwide concerning the impact of fungi on plant disease. With respect to animals, studies show they cause disease primarily only when the animal is in a weakened condition. They are, in essence, **opportunists.**

Figure 19.1 The two major hyphal types found in fungi.

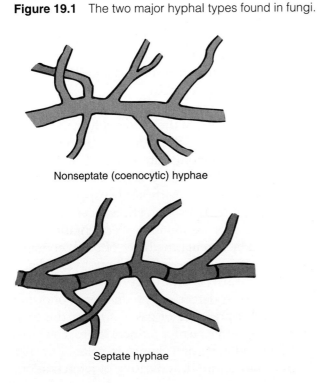

Nonseptate (coenocytic) hyphae

Septate hyphae

Table 19.1   Classification of the Fungi

	Class			
	Zygomycetes	**Ascomycetes**	**Basidiomycetes**	**Deuteromycetes**
Mycelium	Nonseptate	Septate	Septate	Septate
Sexual spores	Oospore (not in a fruiting body) found in aquatic forms; Zygospore (not in a fruiting body) found in terrestrial forms	Ascospores, borne in an ascus, usually contained in a fruiting body	Basidiospores, borne on the outside of a clublike cell (the basidium), often in a fruiting body	None
Asexual spores	Zoospores, motile; Sporangiospores, nonmotile, contained in a sporangium	Conidiospores, nonmotile, formed on the tip of a specialized filament, the conidiophore	Same as Ascomycetes	Same as Ascomycetes
Common representatives	Downy mildews, potato blight, fly fungi, bread mold (*Rhizopus*)	Yeasts, morels, cup fungi, Dutch elm disease, ergot	Mushrooms, puff balls, toadstools, rusts, smuts, stinkhorns	Mostly imperfect Ascomycetes and some imperfect Basidiomycetes*

*Some of these fungi will no doubt form sexual spores in the right environment. In this event, they would need to be reclassified.

Examples illustrating the diverse morphology of yeasts and molds when examined with the microscope are shown in figures 19.2, 19.3, 19.4, and 19.5. The beauty of fungus identification is that they can often be identified to the genus level simply by their macroscopic and microscopic growth characteristics when cultivated on various nutritional media. Most yeasts multiply asexually by a process known as **budding.** An exception is the genus *Schizosaccharomyces*, which has the same vegetative multiplication process as bacteria—**fission** (figure 19.6). A microscopic study of cell and bud morphology can determine if a yeast is a member of the genus *Saccharomyces* or perhaps another genus with a different morphology, for example, *Selenotila* or *Trigonopsis*, (figure 19.2). Also as earlier mentioned, fungi can be easily differentiated from bacteria by their larger size. Yeast and mold identification to the species level often requires additional morphological and physiological studies. Morphologically, some yeasts form sexual spores **(ascospores),** which are borne inside an **ascus.** Examples are *Saccharomyces cerevisiae* (figure 19.7) and *Schizosaccharomyces pombe.*

Some pathogenic yeasts exhibit a property known as yeast **dimorphism.** Examples of dimorphism are shown in table 19.2 under the column titled "Morphology." One of these, the genus *Candida*, forms oval to elongate buds when grown on the surface of Sabouraud's dextrose agar (figure 19.2b), **chlamydospores** and **blastospores** when grown on cornmeal agar (figure 19.8), and **germ tubes** (figure 19.9) when grown in serum or egg albumin.

Physiological tests used for identifying yeast to the species level evaluate the yeast's ability to **assimilate** and/or **ferment** growth media containing various sugars as the sole carbohydrate source. For assimilation studies, agar plates seeded with the test yeast are inoculated on the surface with sterile paper discs containing the various sugars. Growth adjacent to a disc is a positive test for assimilation. Fermentation tests are conducted by inoculating tubes of broth containing different sugars with a drop of the test yeast. Each tube also contains a small inverted glass tube (Durham tube) to detect gas production. Following incubation, the presence of gas (CO_2) in the tube constitutes a positive test for fermentation. The presence of yeast sediment or a change in color of the pH indicator is not, in itself, indicative of fermentation.

Figure 19.2 Drawings of vegetative cell morphology of some representative yeasts. *(a) Saccharomyces cerevisiae,* round to oval cells; *(b) Candida sp.,* elongate to oval cells with buds that elongate, forming a false mycelium; *(c) Selenotila intestinalis,* lenticular cells; *(d) Trigonopsis variablis,* triangular cells.

(a)

(b)

(c)

(d)

Figure 19.3 Intact asexual reproductive structure of the zygomycete *Rhizopus nigricans.* Note nonseptate (coenocytic) hyphae and sporangiophore.

Figure 19.4 Intact asexual reproductive structure of the ascomycete *Aspergillus niger.* Note the presence of a foot cell, a columella, a septate conidiophore, and hyphae.

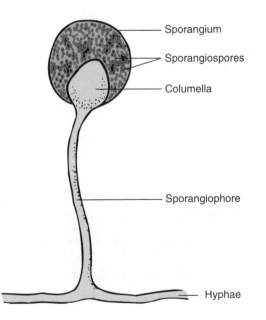

- Sporangium
- Sporangiospores
- Columella
- Sporangiophore
- Hyphae

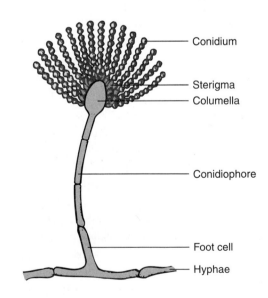

- Conidium
- Sterigma
- Columella
- Conidiophore
- Foot cell
- Hyphae

Figure 19.5 Intact asexual reproductive structure of the ascomycete *Penicillium*. Note the absence of a columella and a foot cell. Also note the symmetrical attachment of the metulae to the conidiophore—an important diagnostic feature for differentiation within the genus. Also note septate hyphae.

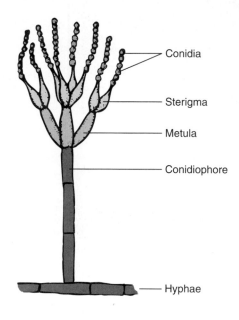

Conidia

Sterigma

Metula

Conidiophore

Hyphae

Figure 19.6 *Schizosaccharomyces pombe,* one of two yeasts known to multiply vegetatively by fission. Courtesy of Dr. David Yarrow, The Central Bureau for Fungus Cultures, Baarn, Holland.

Figure 19.7 Sexual asci and ascospores of *Saccharomyces cerevisiae,* showing asci with four or perhaps fewer spores. Courtesy of University of Washington.

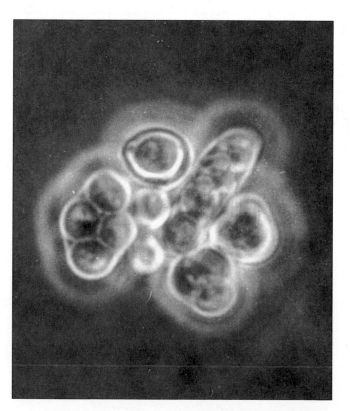

The biochemical pathway for fermentation of sugar into alcohol, CO_2, and other end products, the Embden-Meyerhof pathway **(glycolytic pathway),** is discussed in Nester, et al., p. 132.

The media used for initial isolation of many fungi is malt extract agar for the yeasts and Sabouraud's dextrose agar, with or without antibiotics to inhibit the growth of contaminant bacteria, for the filamentous fungi. The temperature of incubation depends on the organisms sought; 20°–25°C is suitable for most yeasts and filamentous fungi. One exception is *Aspergillus fumigatus*, which grows well at 45°C, a temperature that inhibits growth of most other fungi.

Many of the medically important fungi are found in the classes Ascomycetes and Deuteromycetes (see table 19.1). Most of their infections are opportunist, limited to cutaneous or subcutaneous tissues. Such infections can sometimes progress to systemic involvement with the possibility of death. According to Al-Doory, the

Table 19.2   Some Important Pathogenic Yeasts or Yeastlike (Dimorphic) Organisms

Organism	Morphology	Ecology & Epidemiology	Diseases	Treatment
Cryptococcus neoformans	Single budding cells, encapsulated	Found in soil and pigeons' nests. No transmission between humans and animals. May be opportunistic.	Meningitis, pneumonia, skin infections, visceral organs	Amphotericin B
Candida albicans	Budding cells, pseudomycelium formation chlamydospores	Normal inhabitants of mouth, intestinal tract. Opportunistic infections.	Thrush, vaginitis, nails, eyes, lungs, systemic infections	Alkaline mouth and douche washes; parahydroxy-benzoic acid esters; amphotericin B
Blastomyces dermatitidis	37°C: single large budding cells 20°C: mold with conidia	Disease of North America and Africa. Found occasionally in nature. No transmission between humans and animals.	Primarily lungs. Also skin and bones	High-calorie, high-vitamin diet; bed rest; aromatic diamidines; amphotericin B
Paracoccidioides braziliensis (*Blastomyces braziliensis*)	37°C: single and multiple budding cells 20°C: mold with white aerial mycelium	Confined to South America. Workers in close association with farming.	Chronic granulomatous infection of mucous membranes of mouth, adjacent skin, lymph nodes, viscera	Sulfonamides; amphotericin B
Histoplasma capsulatum	37°C: single small budding cells 20°C: mold with tuberculate chlamydospores	Saprophyte in soil. No transmission between humans and animals. Epidemics from silos, chicken houses, caves, etc.	Primarily lungs, may spread to reticuloendothelial system	Amphotericin B
Coccidioides immitis	37°C: thick-walled, endospore-filled spherical cells 20°C: mold with arthrospores	Disease primarily of arid regions, such as San Joaquin Valley. Dust-borne disease. No transmission between humans and animals.	Primarily lungs, may disseminate particularly in African-Americans. Highly fatal	Bed rest; amphotericin B; surgery for lung lesions

use of new medical technologies, such as prolonged or extensive use of antibiotics, anticancer agents, and immunosuppressive drugs in organ transplants, is expected to continue, thus increasing the ever-present risk of opportunist fungal infections. If so, there will be an expanded need for trained mycologists and clinical mycology laboratories.

Clinically there are three types of these fungal diseases, or **mycoses:** dermatomycoses, subcutaneous mycoses, and systemic yeast and yeastlike infections.

1. **Dermatomycoses** are infections of the skin, hair, and nails caused by a group of filamentous fungi commonly called dermatophytes. They rarely invade subcutaneous tissues. They show rudimentary morphology, appearing only as mycelial growth on skin and nails or as fragments of **mycelium** and **arthrospores** arranged inside and outside of hair. In all instances, they form circular lesions described as **ringworm** (see

Figure 19.8 Chlamydospores and smaller blastospores (attached to pseudohyphae) of *Candida albicans* grown on cornmeal agar. Preparation stained with methylene blue. Courtesy of the Upjohn Co.

Figure 19.9 Germ tubes formed by *Candida albicans* grown on egg albumin. Phase-contrast magnification. Courtesy of the Upjohn Co.

color plate 7). However, in culture they form filamentous colonies and asexual reproductive spores.

2. **Subcutaneous mycoses** are caused by either filamentous or by **dimorphic** yeastlike fungi (see color plate 8). They also remain fixed at the site of infection.

3. **Systemic yeast** and **yeastlike infections** originate in the lungs and can spread to other parts of the body. Examples include **histoplasmosis,** which is caused by the dimorphic yeastlike fungus *Histoplasma capsulatum,* and **coccidioidomycosis** (San Joaquin fever), caused by another

yeastlike fungus, *Coccidioides immitis.* *C. immitis* infection is diagnosed in humans using two widely used serological tests: the ELISA test and the Ouchterlony test. These are discussed in exercises 29 and 30.

Definitions

Arthrospores. Thick-walled asexual spores formed by septate hyphae breaking apart.

Ascospore. Sexual spore characteristic of the fungus class Ascomycetes.

Ascus. Saclike structure containing ascospores.

Assimilation. Ability, in the presence of oxygen, to utilize carbohydrates for growth.

Blastospore. Asexual spore formed by budding from a cell or from hyphae.

Budding. An asexual process of reproduction in which a daughter cell (bud) evolves from either a larger cell (mother cell) or from a hypha.

Chlamydospore. A resistant hyphal cell with a thick wall; it eventually separates from the hyphae and functions as a spore.

Coccidioidomycosis. An upper respiratory tract infection caused by the dimorphic yeastlike organism *Coccidioides immitis.*

Coenocytic. A multinuclear mass of protoplasm resulting from repeated nuclear division unaccompanied by cell division.

Columella. A swelling of the sporangiophore at the base of the sporangium that acts as a support structure for the sporangium and its contents.

Conidiophores. Specialized hyphae that bear asexual spores.

Conidia. Asexual spores produced from either the tip or side of the conidiophore.

Culture spherule. A thick-walled, sphere-shaped cell containing many small endospores, characteristic of the tissue phase of *Coccidioides immitis.*

Conidiospores.   Asexual spores.

Cyst.   A resting (dormant) spore.

Dermatomycosis.   A disease of the skin caused by infection with a fungus.

Dimorphic.   Ability to exist in two forms, *e.g.,* in the fungi, either a mycelial form or a yeastlike form.

Fermentation.   Growth, in the absence of oxygen, in which the final electron acceptor is an organic compound.

Filamentous fungi.   Fungi that grow by hyphae rather than budding.

Fission.   An asexual process in which one cell splits into two or more daughter cells.

Foot cell.   A cell located at the base of the conidiophore in the genus *Aspergillus*.

Germ tube.   A tubelike outgrowth from an asexual yeast cell that develops into a hypha.

Glycolytic pathway.   An initial series of fermentation steps in which carbohydrates are degraded. Often called the Embden-Meyerhof pathway.

Histoplasmosis.   A pulmonary infection caused by *Histoplasma capsulatum*, a dimorphic fungus.

Hypha (pl., hyphae).   Threadlike fungal filament that forms a mycelium.

Lichen.   A symbiotic relationship between a fungus and an alga. See Nester et al. for additional information.

Metula.   A branch(s) at the tip of the conidiophore that supports sterigmata.

Mold.   A filamentous fungus often appearing as woolly growth on decaying materials.

Mycelium.   A fungal mat made of tangled hyphae.

Mycology.   The study of fungi.

Mycoses.   Diseases caused by fungi.

Nonfilamentous fungi.   Fungi devoid of hyphae, *e.g.,* yeast.

Opportunist.   An organism capable of causing disease only when host defense mechanisms are impaired.

Rhizoids.   Rootlike structures made of fungus hyphae that are able to penetrate various substrates in order to anchor the fungus so that it can obtain nutrients.

Ringworm.   Contagious fungal disease of the hair, skin, or nails. See color plate 7.

Saprophyte.   An organism that obtains nourishment from decayed organic matter.

Spherule.   A large thick-walled structure filled with fungal endospores.

Sporangiospore.   Asexual reproductive spore found in the Zygomycetes.

Sporangium.   The sac containing asexual spores.

Sterigma (pl., sterigmata).   A specialized hypha that supports either a conidiospore or a basidiospore.

Stolon.   A runner, such as found in strawberry plants, made of horizontal hyphae from which sporangiospores and rhizoids originate. Stolons are characteristic of the class Zygomycetes.

Subcutaneous mycoses.   Fungal infections which do not spread through the body, but remain localized.

Systemic yeast.   Yeast found in various parts of the body.

Toadstool.   A large filamentous, fleshy fungus with an umbrella-shaped cap.

Yeast.   A nonfilamentous fungus often found in nature on fermenting fruits and grains.

Yeast dimorphism.   Existing in two growth forms, such as the mold phase (hyphal filaments) and yeast phase (single cells) of pathogenic fungi.

Yeast like infections.   Dimorphic fungi that cause such diseases as histoplasmosis and coccidioidomycosis.

Objectives

1. To introduce information about what fungi are and how to distinguish them from one another, as well as how to

identify members of the two major groups of fungi, the nonfilamentous and the filamentous fungi.

2. To introduce information included for identification studies of the nonfilamentous fungi (yeasts) is a member of the class Ascomycetes, *Saccharomyces cerevisiae*. Morphological studies to be used for its identification include colonial and vegetative cell morphology.

3. To introduce information included for studies of the filamentous fungi is a member of the class Zygomycetes, *Rhizopus nigricans*, and two members of the class Deuteromycetes, *Aspergillus niger* and a *Penicillium* species. Morphological studies to be used for their identification are macroscopic and microscopic studies of colonial and vegetative cell morphology when cultured on agar plates.

References

Larone, D. *Medically important fungi, a guide to identification*, 2nd ed. Washington, D.C.: American Soc. for Microbiology, 2002.

Moore, D. *Fungal morphogenesis.* Cambridge University Press, 1998.

Nester et al. *Microbiology: A human perspective*, 6th ed. The McGraw-Hill Companies, 2009. Chapter 12, Section 12.3.

Phaff, H. H.; Miller, M. W.; and Mrak, E. M. *The life of yeasts*, 2nd ed. Cambridge, Mass.: Harvard University Press, 1978. This classic book provides an excellent introduction to yeast morphology, propagation, cytology, ecology (where and how they propagate in nature), and their role as food spoilage organisms, as well as their use in various industries.

Materials

Per team of three students

Cultures of the following:

Sabouraud's dextrose broth cultures (48 hr, 25°C) of *Saccharomyces cerevisiae*

Sabouraud's dextrose agar petri dish cultures (3–5 days, 25°C) of *Rhizopus nigricans, Aspergillus niger,* and *Penicillium notatum.*

Tubes of glucose, maltose, and lactose broth-containing Durham tubes, 2 tubes of each

Glucose-acetate yeast sporulation agar, 1 plate

Cornmeal agar, 1 plate

Sterile droppers, 4

Tweezers

Dissecting microscope(s)

Ruler divided in mm

Dropping bottle containing methylene blue

Procedure

First Session

Suspend the broth culture of *Saccharomyces cerevisiae.*

1. Yeast fermentation study. Inoculate each of the carbohydrate fermentation tubes (glucose, lactose, and maltose) with a loopful of *Saccharomyces cerevisiae.* Place the properly labeled tubes in a container, and incubate at 25°–30°C for 48 hours.

2. Yeast colonial and vegetative cell morphology study. Label a Sabouraud's dextrose agar plate "*S. cerevisiae.*" With a sterile dropper, inoculate the agar surface with a <u>small</u> drop of yeast. Allow the inoculum to soak into the agar before incubating right side up in the 25°–30°C incubator for 48 hours.

3. Yeast sexual sporulation study. With a sterile dropper inoculate the center of the sporulation agar plate with a small drop of the *S. cerevisiae* broth culture. Allow the inoculum to soak into the agar before incubating right side up in the 25°–30°C incubator for 48 or more hours. Cultures freshly isolated from nature generally sporulate much faster than laboratory-held cultures.

4. Colonial characteristics of petri dish cultures of *Rhizopus nigricans, Aspergillus niger,* and *Penicillium notatum* or other species of *Penicillium.*

   Visually examine each petri dish culture noting the following:

   a. Colony size. With a ruler measure the diameter in mm.

b. Colony color. Examine both the upper and lower surfaces.
c. Presence of soluble pigments in the agar medium.
d. Colony texture (such as cottony, powdery, or woolly).
e. Colony edge (margin). Is it regular or irregular?
f. Colony convolutions (ridges). Are they present?

Enter your findings in table 19.5 of the Laboratory Report.

5. Morphological study of asexual fruiting structures found in *Rhizopus*, *Aspergillus*, and *Penicillium* species.

The best way to make studies of this type is with a covered slide culture. In the event none is available, you can attempt to do so with your petri dish culture. A major problem is the density of growth in the petri dish culture, which makes it difficult to find intact asexual reproductive structures. However, they can often be found for *R. nigricans*, in that, like strawberries, it has **stolons,** which enable it to spread and attach to the underside of the petri dish lid. A dissecting microscope is preferred for making initial observations. If not available, the light microscope's low-power objective lens can also be used. The low-power objective lens more than doubles the magnification obtained with the dissecting microscope.

Procedurally do as follows: Place the covered petri dish culture on the stage of either the dissecting microscope or the light microscope, and examine a sparsely populated area of the colony for the presence of asexual reproductive structures (see figure relating to the fungus culture being examined, 19.3, 19.4, or 19.5).

Note: Never smell fungus cultures—spore inhalation may cause infection. When you first observe fruiting bodies, stop moving the petri dish. Keeping the air currents to a minimum to avoid spore dispersal, carefully remove the petri dish cover and re-examine to determine whether you can in fact see the various parts of the fruiting body as described in the figure for that fungus. In the case of *R. nigricans*, you should be able to see fruiting bodies by examining the underside of the petri dish lid. It may take as long as 4–5 days' incubation before finding *Rhizopus* fruiting bodies with stolons. If intact fruiting bodies are found for any of the three filamentous fungi, make drawings of their asexual reproductive structures in part 5 of the Laboratory Report. Label the parts in a manner similar to that used in figures 19.3, 19.4, and 19.5. Covered slide cultures are the answer if fruiting bodies cannot be found for the *A. niger* and *P. notatum*.

6. Detailed examination of **sporangiospores, conidia,** and if present, chlamydospores.

For observing sporangiospores (*R. nigricans*) and conidia (*A. niger* and *P. notatum*), remove some aerial growth with a loop. Prepare a wet mount and observe it with the low-power and high-dry objective lenses of the microscope.

Chlamydospores can be found in both surface and submerged *R. nigricans* mycelium. They are elongated, brown in color and have thick walls. Prepare a wet mount and observe with both the low-power and high-dry objective lenses.

Note: You may wish to first search for spores in both the inner and outer fringes of the colony using the low-power objective lens. To better search the inner colony surface, make the area less dense by first removing some of the aerial growth with a loop. Flame the loop to destroy the spores. In part 6 of the Laboratory Report, prepare and label drawings of the various asexual spore types found.

Some morphological characteristics of value for identification include the following for the three cultures under examination.

1. *Rhizopus nigricans*
   a. Has nonseptate (coenocytic) hyphae.
   b. Contains **rhizoids.** See the underside of the petri dish of an older culture.
   c. Details of fruiting body (figure 19.3): Note the nonseptate stem (sporangiophore), the swelling at the tip of

the sporangiophore **(columella),** and the sac that encloses the columella **(sporangium),** which contains the asexual reproductive spores (sporangiospores).

2. and 3. *Aspergillus niger* and *Penicillium notatum*

These molds have fruiting bodies somewhat similar in appearance. Both have brushlike structures comprising bottle-shaped cells **(sterigmata)** to which are attached long chains of asexual reproductive spores **(conidiospores).** They differ in that the genus *Aspergillus* has a swollen cell at the base of the stem **(conidiophore)** known as the **foot cell,** as well as a columella to which the sterigmata are attached (figure 19.4). The sterigmata may occur in one or two series, depending upon the species involved. Finally, in the genus *Penicillium*, the branching at the tip of the conidiophore can be either symmetrical (figure 19.5) or asymmetrical, depending on where the **metulae** are attached to the conidiophore. When all of the metulae are attached at the tip of the conidiophore, branching of the sterigmata and attached conidia will appear symmetrical (figure 19.5). If one of the metulae in figure 19.5 is attached below the tip of the conidiophore, an asymmetrical branching occurs. This is an important diagnostic feature for differentiation within the genus *Penicillium*.

Second Session

1. Yeast fermentation study. Examine the fermentation tubes for the following and record your results in table 19.3 of the Laboratory Report:
   a. Presence or absence (+ or −) of cloudy broth (growth).
   b. Presence or absence (+ or −) of gas in the inverted Durham tube.
   c. Change in color of the pH indicator dye. A change to a yellow color is indicative of acid production.

Note: Gas production is indicative of fermentation (glycolysis). To detect false-negative results caused by super saturation of the broth, all tubes giving an acid

reaction should be shaken lightly and the cap vented. This operation is frequently followed by a rapid release of gas. All positive fermentation reactions with a carbohydrate sugar are accompanied by positive assimilation of that carbohydrate, evidenced by increased clouding of the broth; however, sugars may be assimilated without being fermented.

2. Yeast colonial and vegetative cell morphology study.
   a. Colony characteristics. If possible, observe the Sabouraud's agar plate over a 5–7 day incubation period. Make note of the following in table 19.4 of the Laboratory Report: colony color; colony consistency (soft, firm—probe the colony with a sterile needle for this determination); colony diameter (mm); colony surface (rough or smooth, flat or raised); and appearance of the colony edge (circular or indented).
   b. Vegetative cell morphology. Remove a loopful of surface growth from each colony and prepare wet mounts. Observe with the high-dry objective lens, noting the shape and size of the cells and the presence or absence of pseudohyphae (figure 19.2b). Prepare and label a drawing of the yeast in part 2b of the Laboratory Report.

3. Sexual sporulation study. With a sterile loop, touch the *S. cerevisiae* colony on the glucose-acetate agar plate and prepare a wet mount. Observe with the high-dry objective lens, looking for the presence of asci containing one to four or perhaps more ascospores (figure 19.7). Prepare and label drawings of your findings in Part 3 of the Laboratory Report.

Note: In the event you do not find asci, reincubate the plate up to one week, perhaps even longer, and reexamine periodically. Some yeast strains take longer than others to produce sexual spores.

4., 5., and 6. Filamentous fungi. Complete any remaining morphological studies.

Note: If an ocular micrometer is available and time permits, you may wish to make measurements of some of the various morphological structures, perhaps a comparison of asexual spore sizes of different filamentous fungi for example. Appendix 5 contains information on use, calibration, and care of the ocular micrometer.

NOTES:

EXERCISE

19

Laboratory Report: Microscopic Identification of Fungi

Results (Nonfilamentous Fungi)

1. Fermentation Study. Examine tubes and record results (+ or −) in table 19.3. For details see Procedure, session 2, step 1, p. 152.

Table 19.3 *Saccharomyces cerevisiae* Fermentation Activity in Tubes of Broth Containing Different Carbohydrate Sugars

Yeast	Glucose			Maltose			Lactose		
	Cloudy	Gas	Acid	Cloudy	Gas	Acid	Cloudy	Gas	Acid
Saccharomyces cerevisiae									

2. Yeast Colonial and Vegetative Cell Morphology Study.
   a. Colony characteristics. For details see Procedure, session 2, step 2a, p. 152. Enter results in table 19.4.

Table 19.4 *Saccharomyces cerevisiae* Colonial Characteristics on Sabouraud's Dextrose Agar Plates

Yeast Strain	Colonial Morphology				
	Colony Color	Consistency	Diameter (mm)	Surface Appearance	Edge Appearance
Saccharomyces cerevisiae					

b. Vegetative cell morphology. For details see Procedure, session 2, step 2*b*, p. 152. Enter results below.

Saccharomyces cerevisiae

3. Sexual Sporulation Study *(Saccharomyces cerevisiae)*. Drawings of asci and ascospores. For details, see Procedure, session 2, step 3, p. 152.

4. Colonial characteristics of *Rhizopus, Aspergillus,* and *Penicillium* when grown on Sabouraud's dextrose agar. Describe in table 19.5.

Table 19.5   Colonial Characteristics of Three Filamentous Fungi Cultured for _____ Days on Sabouraud's Dextrose Agar

	Rhizopus	*Aspergillus*	*Penicillium*
Colony color			
Colony diameter (mm)			
Colony texture			
Colony convolutions			
Colony margin			
Soluble pigments in agar			

5. Drawings of their asexual reproductive structures (please label parts):

       Rhizopus   *Aspergillus*   *Penicillium*

6. Drawings of their asexual spores (please label parts):

       Rhizopus   *Aspergillus*   *Penicillium*

Questions

1. Explain the physiological differences between yeast fermentation and yeast assimilation of glucose.

2. Why is a loop rather than a pipet used to inoculate the sugar fermentation tubes?

3. Why would the growth of a pellicle, or film, on the surface of a broth growth medium be advantageous to the physiology and viability of that yeast?

4. What are some ways in which you might be able to differentiate *Rhizopus nigricans* from *Aspergillus niger* simply by visually observing a petri dish culture?

5. How can you determine whether or not a green, woolly looking colony is *Aspergillus* or *Penicillium*?

6. What problems might you have in identifying a pathogenic fungus observed in a blood specimen? What might you do to correct such problems?

7. In what ways can we readily distinguish
   a. fungi from algae?

   b. fungi from bacteria?

   c. fungi from actinomycetes?

8. Define an opportunistic fungus. Provide some examples. Are all medically important fungi opportunistic? Discuss your answers.

9. Name three pathogenic fungi that exhibit dimorphism. Describe the type of dimorphism each exhibits and the laboratory conditions necessary to elicit it.

NOTES:

Plate 1 Appearance of bacterial colonies growing on agar. © Larry Jensen/Visuals Unlimited.

Plate 2 Appearance of mold (*Penicillium*) growing on agar. © Raymond B. Otero/Visuals Unlimited.

Plate 3 Bacterial shape. *Staphylococcus aureus* cocci as seen with the bright-field light microscope (×1,000). © LeBeau/Biological Photo Service.

Plate 4 Bacterial shape. *Bacillus megaterium* rods as seen with the bright-field light microscope (×600). © George Wilder/Visuals Unlimited.

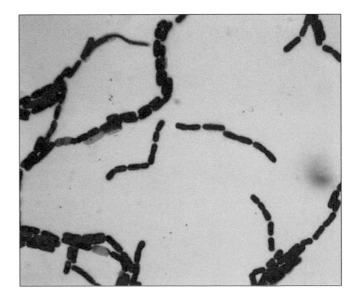

Plate 5 Bacterial shape. *Rhodospirillum rubrum* as seen with the bright-field light microscope (×500). Courtesy of Thomas Tottleben, Tottleben Scientific Company.

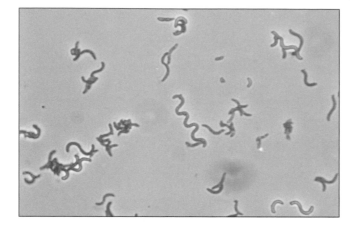

Plate 6 (*a*) Bacteria-producing α-hemolysis on blood agar. (*b*) Bacteria-producing β-hemolysis on blood agar. (*a*) Courtesy of the University of Washington. (*b*) © L.M. Pope & D.R. Grote/Biological Photo Service.

(a)

(b)

Plate 7 Ringworm lesions on the scalp. Note hair loss and scaling of scalp. © Everett S. Beneke/Visuals Unlimited.

Plate 8 Subcutaneous sporotrichosis. Identification is based on the type of lesion formed and culture characteristics. © Everett S. Beneke/Visuals Unlimited.

Plate 9 *Paramecium* from pond water showing cilia and internal structures. Phase-contrast microscopy (×100). © Mike Abbey/Visuals Unlimited.

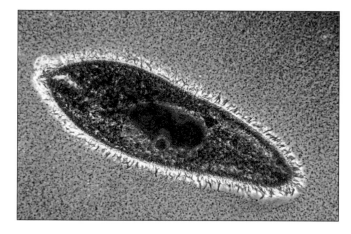

Plate 10 *Schistosoma miracidium,* or snail phase. Note the ciliated larvae. © Cabisco/Visuals Unlimited.

Plate 11 Fermentation results. *From left* (yellow) *to right:* 1. Uninoculated control. 2. No change. 3. Acid and gas. 4. Acid. Courtesy of the University of Washington.

Plate 12 Indol test. *From left to right:* 1. Uninoculated control. 2. Positive for indol formation (red). 3. Negative for indol formation. Courtesy of the University of Washington.

Plate 13 Citrate utilization. *From left to right:* 1. Uninoculated control. 2. No growth, citrate negative. 3. Growth, citrate positive. Courtesy of the University of Washington.

Plate 14 Urease test. *From left to right:* 1. Uninoculated control. 2. Urease negative. 3. Urease positive. Courtesy of the University of Washington.

Plate 15 Rapid bacterial identification system, Enterotube II. Uninoculated tube (*top*) and inoculated tubes (*bottom*). Courtesy of the University of Washington.

Plate 16 Appearance of normal human peripheral blood. Note the frequency of white blood cells as compared to red blood cells. In abnormal blood, the number of white blood cells increases. Copyright © Ed Reschke.

Plate 17 Note line of identity by fusion. See figure 30.3*a* for further explanation. Courtesy of the University of Washington.

Plate 18 Gram-positive cocci growing on mannitol-salt agar plates. *Top: Micrococcus. Bottom left: Staphylococcus aureus. Bottom right: Staphylococcus epidermidis.* Courtesy of the University of Washington.

Plate 19 Light micrograph of Gram-stained. *Escherichia coli.* Courtesy of the University of Washington.

Plate 20 Acid-fast stain of Mycobacterium. The acid-fast bacteria retain the red dye. Courtesy of the University of Washington.

Plate 21 *Clostridium botulinum* endospores stained with malachite green endospore stain. CDC/Courtesy of Larry Stauffer, Oregon State Public Health Laboratory.

Plate 22 The colony on the *right, Pseudomonas aeruginosa,* produces a green pigment. Organism on *left* does not produce a pigment.

Plate 23 Two organisms growing on MacConkey agar. The organism on the *right* is lactose positive (ferments lactose) and the organism on the *left* is lactose negative (does not ferment lactose).

Plate 24 Methyl red test. The tube on the *right* is methyl red positive, and the tube on the *left* is methyl red negative.

Plate 25 Eosin methylene blue (EMB) agar. *Escherichia coli* at *left* is lactose positive and has a metallic green sheen. The organism at *right* is lactose negative.

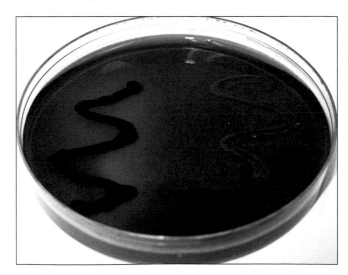

Plate 26 Coagulase test. Organisms in the bottom tube are coagulase positive.

Plate 27 An uninoculated control of Snyder dental caries agar (*left*) and a yellow positive reaction (*right*) at 72 hours, indicative of streptococci from the oral cavity.

Plate 28 Bile esculin slants depicting a positive reaction (*left*) and a negative reaction (*right*).

EXERCISE

20

Parasitology: Protozoa and Helminths

Getting Started

Since the natural histories of parasitic diseases differ in some important respects from those of bacterial diseases, they merit a separate session to give you introductory laboratory experience with parasites, the diseases they cause, and techniques used to diagnose them.

The distinguishing features of parasitic life are the close contact of the parasite with the host in or on which it lives and its dependency on the host for life itself. This special association has led to the evolution of three types of adaptations not found in the free-living relatives of the parasites: loss of competency, special structures, and ecological ingenuity.

Parasites have become so dependent on their hosts for food and habitat that they now experience a *loss of competency* to live independently. They usually require a specific host, and many have lost their sensory and digestive functions; these are no longer important for their survival.

On the other hand, they have developed *special structures* and functions not possessed by their free-living relatives that promote survival within the host. One example is special organs of attachment—hooklets and suckers. Parasites also have a tremendously increased reproductive capacity, which compensates for the uncertainty in finding a new host. Tapeworms, for example, have fantastically high rates of egg production, reaching up to 100,000 per day.

Ecological ingenuity is demonstrated in the fascinating variety of infecting and transmitting mechanisms. This has led to very complex life cycles, which contrast markedly with the relatively simple lifestyles of their free-living counterparts. Parasites show quite a range in the types of life cycles they possess, from species that pass part of each generation in the free-living state to others that require at least three different hosts to complete the life cycle. Some are simply transmitted by insects from

one human host to the new host, or the insect may act as a host as well. Many protozoa develop resistant cysts, which enable them to survive in unfavorable environments until they find a new host. The eggs of flatworms and roundworms also have a protective coat.

These three strategies promote survival and expansion of the species by providing greater opportunities for finding and infecting new hosts, which is a continual problem for parasites. Successful interruption of these cycles to prevent their completion is an important feature of public health measures used to control diseases caused by parasites.

This exercise is designed to give you some practical experience with representative protozoan and helminthic parasites and with clinical methods used in their diagnosis and control. Your study will consist of these procedures:

1. As an introduction, you will have an opportunity to observe the movements and structure of some living nonparasitic protozoans and worms often found in pond water.
2. Examination of commercially prepared stained blood and fecal slides that contain human protozoan parasites.
3. Microscopic comparison of the structure of parasitic worms with that of their free-living relatives to observe some special adaptations to the parasitic way of life.
4. Study of the natural history and life cycle of the human parasitic disease schistosomiasis. This will enable you to see the interaction between stages of the life cycle, environmental surroundings, and social conditions of their human hosts as factors in the epidemiology and control of the disease.

The following classification of parasites will serve as a guide to the examples you will be studying in this exercise. It is not a complete listing.

Protozoa

Protozoa, a subkingdom of the kingdom Protista, are unicellular eukaryotic organisms. They usually reproduce by cell division and are classified mainly according to their means of locomotion. Only one phyla, the Suctoria, which is closely related to the Ciliophora, does not contain animal pathogens. The remaining are classified as follows:

Sarcodina

Members of this subphylum move and feed slowly by forming cytoplasmic projections known as **pseudopodia** (false feet). They also form both **trophozoites** (vegetative form) and **cysts** (resistant, resting cells). Parasitic members include the **amoeba** *Entamoeba histolytica*, which causes amoebic dysentery. It ingests red blood cells and forms a four-nucleate cyst. It is also found in animals. Other amoeba species found in humans, such as *Entamoeba gingivalis*, are relatively harmless **commensals.**

Ciliophora

Members of this phylum have many short, hairlike cilia on their body surfaces that beat rhythmically by bending to one side. They contain two nuclei: a macronucleus and a micronucleus. Ciliophora is typified by the genera *Paramecium* (see color plate 9) and *Stentor*, both nonpathogens readily found in pond water. Another member, *Balantidium coli* (figures 20.1a, 20.1b), is a common parasite in swine. It possesses both a cyst and trophozoite form and can infect humans, causing serious results.

Mastigophora

These protozoans propel themselves with one or more long, whiplike flagella. Some have more than one nucleus, and they usually produce cysts. Different species cause infections in the intestines, vagina, blood, and tissues. *Giardia lamblia* (see figure 32.3) causes a mild to severe diarrheal infection. *Trichomonas vaginalis* (figure 20.2) is found in the urogenital region, where it causes a mild vaginitis in women. *Trypanosoma gambiense* (figure 20.3) infects the blood via tsetse fly bites, causing trypanosomiasis, or African sleeping sickness, in cattle and humans. Cattle and other ungulates serve as a reservoir for this organism.

Figure 20.1 (a) *Balantidium coli* cysts viewed on 40 x. The macronucleus can be seen on the photo on the right. (b) *Balantidium coli* trophozoite viewed on 40 x, showing the nucleus overlaid on the macronucleus.

(a)

(b)

Sporozoa

Sporozoa are obligate, nonmotile parasites with alternating stages: the sexual reproductive stage is passed in the definitive insect host, and the asexual phase is passed in the intermediate human or animal host. The genus *Plasmodium* includes the malarial species, in which the **definitive host** is the female *Anopheles* mosquito and the **intermediate host** is humans. The genus *Coccidia* includes important intestinal parasites affecting fowl, cats, dogs, swine, sheep, and cattle. *Toxoplasma gondii* is a cat parasite that can harm the human fetus in an infected pregnant woman.

Helminths (Worms)

Helminths are multicellular eukaryotic organisms. Two of the phyla, Platyhelminthes (flatworms) and Nematoda (roundworms), contain pathogenic worms.

Figure 20.2 (*a*) *Trichomonas vaginalis.* Illustration of a typical mastigophoran protozoan. (*b*) *Trichomonas vaginalis* viewed on 100 x. The flagella, nucleus, axostyle, and undulating membrane can all be seen on the right.

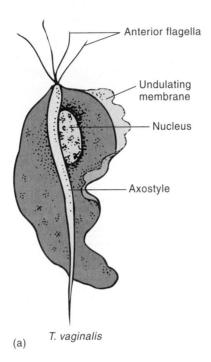

Anterior flagella

Undulating membrane

Nucleus

Axostyle

T. vaginalis

(a)

(b)

Figure 20.3 A prepared blood smear of trypanosome parasites in human blood. The slender flagellates lie between the red blood cells. © Cleveland P. Hickman, Jr./Visuals Unlimited

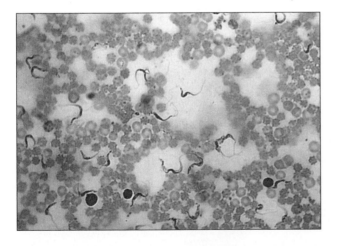

Phylum Platyhelminthes

Members of Platyhelminthes are flat, elongated, legless worms that are **acoelomate** and exhibit bilateral symmetry. This phylum contains three classes:

Class Turbellaria

Members of this class are free-living **planarians** (flatworms), such as are found in the genera *Dugesia* (figure 20.4) and *Planaria*.

Class Trematoda (Flukes)

Flukes have an unsegmented body, and many have suckers to hold them onto the host's intestinal wall. Many flukes have complex life cycles that require aquatic animal hosts. The *Schistosoma* species are bisexual trematodes that cause serious human disease. They require polluted water, snails, and contact with human skin for completion of their life cycles (see color plate 10). *Clonorchis sinensis* and *Fasciola* species are liver flukes acquired by eating infected raw fish and contaminated vegetables, respectively.

Class Cestoda (Tapeworms)

Tapeworms are long, segmented worms with a small head (**scolex**) equipped with suckers and often hooklets (figure 20.5) for attachment to the host's intestinal wall. The series of segments, or **proglottids,** contain the reproductive organs and thousands of eggs. These segments break off and are eliminated in the feces, leaving the attached scolex to produce more proglottids with more eggs. Figure 20.6 illustrates the life cycle of the tapeworm in a human. The symptoms of *Taenia* tapeworm infection are usually not serious, causing only mild intestinal symptoms and loss of nutrition. Not so for the *Echinococcus* tapeworm, which causes a serious disease. All tapeworm diseases are transmitted by animals.

Phylum Nematoda

Members of Nematoda (roundworms) occupy an important ecological niche since they are present in large numbers in very diverse environments, including soil, fresh water, and seawater. In contrast to the Platyhelminthes, these round, unsegmented worms are **coelomate** (have a body cavity), and have a complete digestive tract and separate sexes. This phylum contains many agents of animal, plant, and human parasitic diseases. Most require only one host and can pass part of their life cycle as free-living larvae in the soil. *Trichinella spiralis* requires alternate vertebrate hosts. Humans become infected when they ingest inadequately cooked meat, such as pork or bear containing the larval forms in the muscles. *Ascaris lumbricoides* (figure 20.7) is probably the most common worldwide of all the human helminths. *Enterobius vermicularis* causes pinworm, a very common condition

---

Figure 20.4   The genus *Dugesia,* a free-living planarian in the class Turbellaria. *(a)* A living specimen (× 0.5), and *(b)* a labeled line drawing. *(a)* © John D. Cunningham/Visuals Unlimited

(a)

Eyespots

Brain

Nerve cord

Pharynx

Mouth

Gastrovascular cavity

(b)

Figure 20.5 Illustration of a tapeworm scolex showing both hooklets and suckers for attachment to the intestine. *Taenia saginata* (beef tapeworm) is essentially without hooklets, whereas *Taenia solium* (pork tapeworm) has both.

© Stanley Flegler/Visuals Unlimited

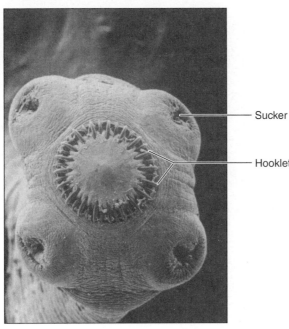

— Sucker

— Hooklets

in children in the United States. Efforts to eradicate it have not been very successful since pinworm causes little discomfort. *Oxyuris* causes a similar condition in animals. *Dilofilaria immitis* (figure 20.8) is transmitted via mosquitoes and causes heartworms in animals. Left untreated, or treated with high doses of Ivermectin, animals will perish.

Definitions

Acoelomate.   Without a true body cavity. Typical of members of the Phylum Platyhelminthes (flatworms).

Amoeba.   Unicellular organisms with an indefinite changeable form.

Cercaria.   The last miracidium stage in which the larvae possess a tail.

Coelomate.   With a true body cavity. Typical of members of the Phylum Nematoda (roundworms).

Figure 20.6   Life cycle of *Taenia saginata*. The adult tapeworm with scolex and proglottids is conceived from larvae in the human intestine.

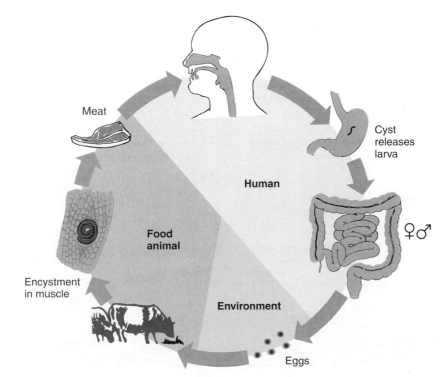

Meat

Cyst releases larva

Human

♀♂

Food animal

Environment

Encystment in muscle

Eggs

Figure 20.7 *Ascaris lumbricoides,* an intestinal roundworm up to 12 inches long. (*a*) A mass of worms recovered from the ileum of a malnourished child. (*b*) *Ascaris lumbricoides* eggs. (*a*) From Rubin and Farber, *Pathology.* Reprinted by permission of J. B. Lippincott Company.

(a)

(b)

Figure 20.8 A prepared blood smear viewed on 40 x showing *Dilofilaria immitis.*

Commensal.   A relationship between two organisms in which one partner benefits from the association and the other is unaffected.

Cysts.   Dormant, thick-walled vegetative cells.

Definitive host.   The host in which the sexual reproduction of a parasite takes place.

Intermediate host.   The host that is normally used by a parasite in the course of its life cycle and in which it multiplies asexually but not sexually.

Merozoites.   Schizont nuclei that become surrounded by cytoplasm and bud off as daughter cells or merozoites.

Miracidium.   A free-swimming ciliate larva that seeks out and penetrates a suitable intermediate snail host, in which it develops into a sporocyst.

Planarian.   Any flatworm of the genus *Planaria*.

Proglottid.   Any of the segments of a tapeworm formed in the neck region by a process of strobilation (transverse fission).

Pseudopodia.   Extensions of cytoplasm that aid in engulfing particles and functioning in motility of amoeboid cells.

Schizont.   A stage in the life cycle of Sporozoa in which the nucleus undergoes repeated nuclear division without corresponding cell divisions.

Scolex.   The head of a tapeworm, which is used for attaching to the host's intestinal wall.

Sporocyst.   A stage in the life cycle of certain protozoa in which two or more of the parasites are enclosed within a common wall.

Trophozoites.   Vegetative forms of some protozoans.

Objectives

1. To introduce you to the study of parasitology by letting you examine some examples of nonparasitic protozoans (*Amoeba proteus* and a *Paramecium* species) and a free-living planarian (flatworm) of the genus *Dugesia*.

2. To study the morphology of some free-living, trophozoite, and cystic forms of intestinal parasites using prepared slides, and malarial and trypanosome parasites using stained blood smears (figure 20.9).

3. To study the natural history and life cycle of an important human parasitic disease, schistosomiasis, using stained slides and a life cycle diagram.

4. To demonstrate special adaptations of parasitic worms through a study of stained slides in which they are compared to similar free-living relatives.

References

Leventhal, R., and Cheadle, R. *Medical parasitology*, 4th ed. Philadelphia: F. A. Davis, 1994.

Markell, E.; Voge, M.; and John, D. *Medical parasitology*, 8th ed. Philadelphia: Saunders, 1999.

Nester et al. *Microbiology: A human perspective*, 6th ed. The McGraw-Hill Companies, 2009. Chapter 12, Section 12.2 and Section 12.5.

Figure 20.9   *Plasmodium falciparum* infection showing the ring trophozoite stage in a blood smear. Courtesy of the Centers for Disease Control.

Ring trophozoite

Neva, F. A., and Brown, H. W. *Basic clinical parasitology*, 6th ed. Norwalk: Appleton Lange, 1994.

Roberts, L. S., and Janovy, J. *Gerald D. Schmidt Larry S. Robert's Foundations of Parasitology*, 6th ed. The McGraw-Hill Companies, 2000.

Materials

Cultures of the following:

Living cultures of a *Paramecium* species, *Amoeba proteus*, and a *Dugesia* or *Planaria* species

If available, a fresh sample of quiescent, stagnant pond water, which often contains members of the above genera. Students may wish to bring their own pond water.

Planaslo solution, 1 or more dropping bottles

Depression slides (hanging drop slides)

The following commercially prepared slides:

Subkingdom Protozoa

Sarcodina (pseudopodia)

Entamoeba histolytica trophozoite and cyst stages

Ciliophora (cilia)

Paramecium trophozoite or *Stentor*

Balantidium coli

Mastigophora (flagella)

Giardia lamblia trophozoite and cyst stages

Trichomonas vaginalis

Trypanosoma gambiense

Sporozoa (nonmotile)

Plasmodium vivax or *P. falciparum* ring, amoeboid

schizont stages

Subkingdom Helminths (worms)

Phylum Platyhelminthes (flatworms)

Class Turbellaria (free living)

Dugesia or *Planaria* species

Class Trematoda (flukes)

Schistosoma mansoni

adult male

adult female

ovum (egg)

ciliated miracidium

infective ciliate cercaria

sporocyst stage in snail liver tissue

Clonorchis sinensis

Class Cestoda (tapeworms)

Taenia solium trophozoite or *T. saginata*

Class Nematoda

Ascaris lumbricoides

Dilofilaria immitis

Procedure

Note: If the number of prepared slides is limited, these procedures may be performed in a different order to facilitate sharing.

1. Examination of free-living cultures.
   a. Pond water examination. Prepare a wet mount (see exercise 3) using petroleum jelly to adhere a coverslip to a glass slide. Examine initially with the low-power objective lens and later with the high-dry and oil immersion objective lenses. Observe the mode of locomotion of any amoeboid or paramecium-like protozoans found. If their movements are too rapid, add a drop of Planaslo to slow them down. Describe their movements and prepare drawings in part 1*a* of the Laboratory Report.
   b. Examination of fresh samples of an amoeba (such as *Amoeba proteus*), paramecium (for example, *Paramecium caudatum*), and a free-living flatworm (such as *Dugesia* or *Planaria* species). Use wet mount slide preparations and examine as described in 1*a* for pond water. Record your observations in part 1*b* of the Laboratory Report.

2. Examination of stained slides for trophozoites and cysts.
   a. Using the oil immersion objective lens, examine prepared slides of a protozoan, either the amoeba *Entamoeba histolytica* or the flagellate *Giardia lamblia*. In the trophozoite stage, observe the size, shape, number of nuclei, and presence of flagella or pseudopodia. In the cyst stage, look for an increased number of nuclei and the thickened cyst wall.
   b. Sketch an example of each stage, label, and record in part 2 of the Laboratory Report.
3. Examination of protozoans present in stained blood slides.
   a. Use the oil immersion objective lens to explore a smear of blood infected with *Plasmodium vivax*, and locate blood cells containing the parasite. After a mosquito bite, the parasites are carried to the liver, where they develop into **merozoites.** Later, they penetrate into the blood and invade the red blood cells, where they go through several stages of development. The stages are the delicate ring stage (figure 20.9), the mature amoeboid form, and the **schizont** stage, in which the organism has divided into many individual infective segments that will then cause the red blood cell to rupture, releasing the parasites, which can then infect other cells. Sketch the red blood cells with the infective organism inside them, and note any changes in the red blood cell shape, pigmentation, or the presence of granules due to the effect of the parasite. Identify and label the stage or stages seen, as well as the species, in part 3a of the Laboratory Report.
   b. Examine the trypanosome blood smear (figure 20.3) with the oil immersion lens, and locate the slender flagellates between the red blood cells, noting the flagellum and undulating membrane. Sketch a few red blood cells along with a flagellate in part 3b of the Laboratory Report.
4. Comparison of a free-living worm with its parasitic relative.
   a. Observe a prepared slide of a free-living flatworm (*Dugesia* or *Planaria* species) with the low-power objective lens. Note the pharynx, digestive system, sensory lobes in the head region, and the eyespots (figure 20.4). Next examine a parasitic fluke such as *Clonorchis*. Note the internal structure, especially the reproductive system and eggs if female, and the organs of attachment, such as hooklets or round suckers.
   b. Sketch each organism in part 4a of the Laboratory Report, and label the main features of each. Describe the main differences between the fluke and the free-living planaria.
   c. Examine prepared slides of a tapeworm (*Taenia* species, figure 20.5), observing the small head, or scolex, and the attachment organs—the hooklets or suckers. Then locate the maturing proglottids along the worm's length. The smaller proglottids may show the sex organs better; a fully developed proglottid shows the enlarged uterus filled with eggs. Sketch, label, and describe its special adaptations to parasitic life in part 4b of the Laboratory Report.
   d. Observe a prepared slide of *Ascaris lumbricoides*, noticing differences between the flatworm, fluke, and tapeworm. Sketch in part 4c of the Laboratory Report.
   e. Observe a prepared slide of *Dilofilaria immitis* on 40X, noting the differences between the other classes of worms. Sketch in part 4d of the Laboratory Report.
5. Life cyle of *Schistosoma mansoni* and its importance in the control of schistosomiasis.
   a. Assemble five or six slides showing the various stages in the schistosoma life cycle: adult worm (male and female if available), ova, ciliated miracidium, the sporocyst in the snail tissue, and the infective ciliate cercaria.
   b. Next read this brief summary of the natural history of *Schistosoma mansoni* (see color plate 10 and **Nester et al. Microbiology: A human perspective 6th ed. The McGraw-Hill Companies, 2009, pp. 298–299**).

   Human schistosomiasis occurs wherever these conditions exist: water is polluted with human wastes; this water is used for human bathing and wading, or irrigation of cropland; and the presence of snail species that are necessary as hosts for the sporocyst stage in fluke development and completion of its life cycle. Solution to this public health problem is very complex, not only because of technical difficulties in its control and treatment, but also because its life cycle presents an ecological dilemma. Many developing countries need food desperately, but the main sources now available for these expanding needs are fertile deserts, which have adequate nutrients but require vast irrigation schemes, such as the Aswan Dam in Egypt. However, due to the unsanitary conditions and the presence of suitable snail hosts, these projects are accompanied by an increase in the disease schistosomiasis, which currently is very difficult to control and very expensive to treat on a wide scale.

The **cercaria** larvae swim in the contaminated water, penetrating the skin of agricultural workers who are barefoot. They migrate into the blood and collect in the veins leading to the liver. The *adults* develop there, copulate, and release the *eggs*. The eggs are finally deposited in the small veins of the large intestine, where their spines cause damage to host blood vessels. Some eggs die; however, others escape the blood vessels into the intestine and pass with the feces into soil and water. There they develop, and then hatch into motile **miracidia,** which eventually infect suitable snail hosts and develop into saclike **sporocysts** in the snail tissues. From this stage, the fork-tailed cercaria larvae develop; these leave the snail and swim in the water until they die or find a suitable human host, thus completing the complex life cycle involving two hosts and five separate stages.

c. Now look at the prepared slides of all the schistosoma stages discussed in the preceding description. Sketch each stage in the appropriate place in the life cycle diagram shown in part 5 of the Laboratory Report.

EXERCISE

20

Laboratory Report: Parasitology:
Protozoa and Helminths

Results

1. Examination of Free-Living Cultures
    a. Pond water examination.
        Description of movements and drawings of any protozoans found in pond water.

    b. Examination of fresh samples of a free-living amoeba, paramecium, and flatworm. Description of
        movements and drawings with labels.

2. Examination of Stained Slides for Trophozoites and Cysts
   Prepare drawings of the trophozoite and cyst stage of either *Entamoeba histolytica*, *Giardia lamblia*, or *Balantidium coli*. Label accordingly (see Procedure step 2).

3. Examination of Protozoans Present in Stained Blood Slides
   a. Examine blood smears of *Plasmodium vivax* (see Procedure step 3a).

   b. Blood smear of *Trypanosoma gambiense* (see Procedure step 3b).

4. Comparison of a Free-Living Worm with Its Parasitic Relative
   a. Comparison of *Dugesia* or *Planaria* species (free-living) with *Clonorchis sinensis* (parasitic). See Procedure step 4a.

b. Study of a parasitic tapeworm (*Taenia* species). See Procedure step 4c.

c. Study of *Ascaris lumbricoides*. See Procedure step 4d.

d. Study of *Dilofilaria immitis*. See Procedure step 4e.

5. Life Cycle of *Schistosoma mansoni* and Possible Methods of Control

   a. For each space in this life cycle, sketch the appropriate stage, using the prepared microscope slides.

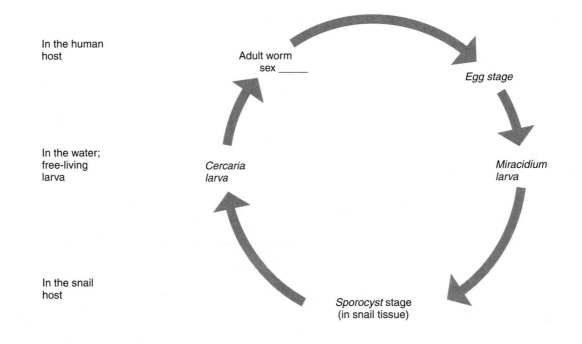

In the human
host

Adult worm
sex _____

Egg stage

In the water;
free-living
larva

*Cercaria
larva*

*Miracidium
larva*

In the snail
host

Sporocyst stage
(in snail tissue)

   b. Propose a plan for public health control of schistosomiasis. Describe various strategies that might be developed by public health personnel to interrupt this cycle and thus prevent schistosomiasis. Show on a diagram where specific measures might be taken, and label. Explain each possibility and its advantages and disadvantages.

Questions

1. Which form—the trophozoite or the cyst—is most infective when found in a feces sample? Explain.

2. In what ways are free-living and parasitic worms similar, such that they can be identified as closely related?

3. In what ways do the parasitic species differ from the free-living planaria? Use the chart to summarize your comparisons.

	Planaria	Fluke	Tapeworm
Outside covering			
Organs of attachment			
Sensory organs			
Digestive system			
Reproduction			

4. Estimate the length and width of a trypanosome. See figure 20.3 for a clue. Show your calculations.

5. How is the *Echinococcus* tapeworm transmitted to humans? Does it cause a serious disease? What are two ways in which its transmission to humans can be prevented?

NOTES:

EXERCISE 21

Titering Prokaryotic Viruses

Getting Started

Viruses cannot replicate without infecting a living cell. To study a virus, you must also grow its host, and so the easiest viruses to work with are those that infect bacteria. These are called **bacteriophage,** or simply phage (rhymes with rage). There are many kinds of bacteriophage, but this discussion is limited to DNA phage that have been well studied, such as lambda and T4. These phage first attach to the bacterial cell wall and then inject their DNA into the cytoplasm. There are two major outcomes of this injection, depending on whether the phage are lytic or lysogenic.

1. **Lytic.** The cell lyses about 30 minutes after infection, releasing approximately 100 virus progeny (figure 21.1).
2. **Lysogenic** (or temperate). The DNA of the phage integrates into the bacterial chromosome and is replicated with the bacterial DNA. It may at some time in the future leave the chromosome when it will direct the production of virus and lyse the cell. Bacteria that contain the DNA of a phage cannot be reinfected or lysed by the same type of phage. The lysogenized cell has the advantage of being

Figure 21.1 Steps in the replication of a T-even phage during the infection of *E. coli.*

(a) The phage attach to specific receptors on the cell wall of *E. coli.*

(b) Following adsorption, phage DNA is injected into the bacterial cell, leaving the phage coat outside.

DNA

(c) Phage mRNA is transcribed from phage DNA.

mRNA

Phage-induced proteins

(d) Phage coat proteins, other protein components, and DNA are synthesized separately.

(e) Phage components are assembled into mature virions—maturation.

(f) The bacterial cell lyses and releases many infective phage.

immune to a second infection by the same phage, but it has the disadvantage of carrying the DNA of a virus that can lyse the cell if it is excised from the cell's chromosome.

Phage are too small (about 200 nm) to be seen in a light microscope, but they can be detected if grown on a bacterial lawn as follows. Phage and their host cells are mixed in a small tube of soft agar and then poured on top of an agar base plate. (Soft agar contains about half the concentration of standard agar so that the phage can diffuse more easily.) The plates are then incubated overnight at the optimum growth temperature for the host bacteria.

During incubation, bacteria multiply and produce a thick covering of bacteria, or **bacterial lawn,** except in those places where phage have infected and killed the bacteria, leaving clear areas called **plaques** (figure 21.2). Since each plaque originated with one phage, the plaques can be counted—just as bacterial colonies—to determine the number of phage originally mixed with the soft agar.

Although the appearance of the plaques can be influenced by many factors, in general **virulent phage** produce clear plaques. **Temperate phage** produce cloudy plaques because many cells within the plaque were lysogenized instead of lysed and thus continue to grow and multiply. The plaques do not

increase in size indefinitely because phage can replicate only in multiplying bacteria. The bacteria and the phage usually have stopped growing after 24 hours.

It is important to study phage and to learn the techniques used to manipulate them for the following reasons.

1. Animal viruses, including human pathogens, are grown on tissue culture cells in the same fashion as phage on bacteria. Tissue culture cells are animal cells that are grown in bottles and plates. The animal virus can form plaques by causing cells to degenerate or die. Tissue culture cells require a more complex (and more expensive) medium, so it is convenient to learn viral technique with phage and bacteria.

2. Phage are used in recombinant DNA experiments and are also useful in studying the genetics of bacteria.

3. They are used to identify different strains of bacteria because one type of phage will only infect a few specific strains.

4. Lysogeny has served as a model for viruses inserting their DNA in animal cells. The life cycle of temperate phage resembles animal retroviruses, such as the HIV virus, the causative agent of AIDS.

If you were to isolate phage that infect *Escherichia coli* from nature, sewage would be a good source. It contains high numbers of *E. coli* and therefore a high concentration of *E. coli* phage. It is possible to filter samples of sewage to obtain just a filtrate of virus. The bacteria are removed by the filter, and the virus particles are small enough to pass through the filter. This is interesting to do, but the filtrate also contains animal virus that could be a health hazard.

In this exercise, you will **serially dilute** a suspension of phage so that you can count an appropriate number of plaques on a plate and calculate the titer of phage. The number of phage/ml is the **titer.** The phage producing a plaque are also called **plaque-forming units.** Sometimes phage are present but for some reason do not form a plaque.

Definitions

Bacterial lawn. The confluent growth of bacteria on an agar plate; also called lawn of bacteria.

Bacteriophage. A virus that infects bacteria; often abbreviated phage (rhymes with rage).

Lysogen. A bacterium carrying a phage integrated in its chromosome. See *temperate phage*.

Lytic virus. A virus that replicates within a host cell and causes it to produce phage, rupture, and die. See *virulent phage*.

Plaque. A clear or cloudy area in a bacterial lawn caused by phage infecting and lysing bacteria.

Plaque-forming units. The single phage that initiates the formation of a plaque.

Serial dilution. A dilution of a dilution, continuing until the desired final concentration is reached.

Temperate phage. A phage that can either integrate into the host cell DNA or replicate outside the host chromosome, producing more phage and lysing the cell. See *lysogen*.

Titer. The concentration of virus in a sample (number/volume) (when used as a noun), or to determine the concentration (when used as a verb).

Virulent phage. A phage that always causes lysis of the cell following phage replication. Unlike a temperate phage, it cannot be integrated into the chromosome of the host. See *lytic virus*.

Objectives

1. To isolate a phage from its natural habitat.
2. To titer a phage suspension.
3. To gain an appreciation of the phage life cycle (to view steps in a phage life cycle).

Reference

Nester et al. *Microbiology: A human perspective*, 6th ed. The McGraw-Hill Companies, 2009. Chapter 13.

Titering a Phage Suspension

Materials

Host bacteria for phage in late log phase
Per team
 Escherichia coli
 phage T4 suspension
 9-ml tryptone blanks, 4
 4-ml overlay agar tubes, 4
 Tryptone agar base plates, 4
 Sterile 1-ml pipets, 5

Procedure

First Session

1. Label four 9-ml tryptone blanks: 10^{-1}, 10^{-2}, 10^{-3}, 10^{-4} (figure 21.3).
2. Transfer 1 ml of the bacteriophage to the tube labeled 10^{-1} with a sterile 1-ml pipet. Discard the pipet. You must use fresh pipets each time so that you do not carry over any of the more concentrated phage to the next dilution.
3. Mix and transfer 1 ml of the 10^{-1} dilution to the 10^{-2} tube and discard the pipet.
4. Mix and transfer 1 ml to the 10^{-3} tube and discard the pipet.
5. Mix and transfer 1 ml to the 10^{-4} tube and discard the pipet.
6. Label four tryptone hard agar base plates: 10^{-1}, 10^{-2}, 10^{-3}, 10^{-4}.
7. Melt four tubes of soft overlay agar and place in a 50°C water bath. Let cool for about 10 minutes.
8. Add about 0.1 ml (or several drops) of *E. coli* broth to each tube of melted overlay agar.
9. Starting with the most diluted phage tube (10^{-4}), add 1 ml to the overlay agar and immediately pour on the tryptone agar base plate labeled 10^{-4}.

Figure 21.3 Schematic showing the procedure used to demonstrate *E. coli* phage plaques on the surface of agar plates. Plaques are represented by the light-colored areas on the 10⁻³ and 10⁻⁴ dilutions. (These results show only one possibility.)

21–4   Exercise 21   Titering Prokaryotic Viruses

10. Using the same pipet, add 1 ml of the 10^{-3} dilution to a tube of overlay agar and pour into the plate labeled 10^{-3}. You can use the same pipet because you started with the most dilute sample and therefore the phage carried over are minimal.
11. Repeat for the 10^{-2} and 10^{-1} phage dilution.
12. Incubate the plates inverted at 37°C after the agar has hardened.

Second Session

1. Examine the plates. Select a plate containing between 30 and 300 plaques. As you count the plaques, place a dot with a marking pen under each plaque on the bottom of the petri plates. These marks can be wiped off so that each team member can count the plaques.
2. Estimate the numbers on the other plates. They should vary by a factor of 10 as the dilution increases or decreases.
3. To determine the titer, use this formula:

No. of plaques × 1/dilution × 1/ml of sample =
plaque forming units/ml.

Example: If 76 plaques were counted on the 10^{-4} dilution, then:

$$76 \times 1/10^{-4} \times 1/1 = 76 \times 10^4 \text{ pfu/ml.}$$

4. Record results.

NOTES:

EXERCISE

21

Laboratory Report: Titering Prokaryotic Viruses

Results

1. Titering a Phage Suspension

Dilution	Control	10^{-1}	10^{-2}	10^{-3}	10^{-4}
number of plaques					

    a. Which dilution resulted in a countable plate?

    b. Did the number of plaques decrease 10-fold with each dilution?

    c. How many phage/ml were in the original suspension? Show calculations (see exercise 8).

Questions

1. How would an agar plate appear after incubation

    a. if you forgot to add the bacteria to the phage/bacteria mixture?

    b. if you forgot to add the phage to the phage/bacteria mixture?

2. How can you distinguish a lytic phage from a temperate phage by observing their plaques? Which did you titer?

3. Why can a plaque be considered similar to a bacterial colony?

4. If you continue to incubate the plates, will the size of the plaques increase?

*I*NTRODUCTION to Medical Microbiology

It is easy to think of microorganisms as a deadly, vicious force—especially when the diseases they cause kill young people or wipe out whole populations. The organisms, however, are simply growing in an environment they find favorable.

If a pathogen becomes too efficient at taking advantage of its host, the host dies and the organism dies with it. Thus, the most successful pathogens are those that live in balance with their hosts. When a new pathogen enters the population, it is very virulent, but after awhile there is a selection toward less virulent pathogens and also a selection in the hosts for increased resistance.

Medical microbiology continues to offer challenges to those interested in medicine and in pathogenic bacteria. These next exercises are an introduction to many of the organisms that are encountered in a clinical laboratory. Not only will you study the characteristics of the organisms, but also you will learn some strategies for isolating and identifying them. In addition, these exercises are designed to help you learn to differentiate between organisms you can expect to find as normal biota in various places in the body and others that are responsible for certain diseases.

NOTES:

EXERCISE

22

Normal Skin Biota

Getting Started

The organisms growing on the surfaces and in the orifices of the body are called normal biota. They are usually considered **commensals** because they do not harm their hosts; in fact, they have several beneficial roles. Normal biota prevent harmful organisms from colonizing the skin because they are already established there and utilize the available nutrients. Some produce enzymes or other substances that inhibit nonresident organisms. Other organisms, called transients, can also be found on the skin for short periods, but they cannot grow there and soon disappear.

Familiarity with organisms making up the skin biota is useful because these organisms are frequently seen as contaminants. Skin is continually flaking off, and bacteria floating in the air on rafts of skin cells sometimes settle into open petri dishes. If you are familiar with the appearance of *Staphylococcus* and *Micrococcus* colonies, for instance, you will suspect contamination if you see such colonies on an agar plate. *Staphylococcus epidermidis* can also be seen in clinical specimens such as urine samples (figure 22.1). These organisms probably are not causing disease but are simply contaminants from skin biota.

Some of the organisms you may isolate include the following:

Staphylococcus epidermidis This Gram-positive coccus is found on the skin as part of the normal biota of almost all humans throughout the world. It can also be isolated from many animals.

Staphylococcus aureus At least 20% of the population carry (have as part of their normal biota) this bacterium. It is found on the skin, especially in the nares, or nostrils, and it seems to cause no harm to its host. However, *S. aureus* is frequently the cause of wound infections and food poisoning, and it has been implicated as the cause of toxic shock syndrome. It can be identified by the **coagulase test.** Recently, many additional species of *Staphylococcus* have been identified. They are associated mostly with diseases in immunologically compromised individuals.

Micrococcus luteus This is a Gram-positive coccus found on the skin of some people, but it almost never causes disease. It is frequently an air contaminant forming bright yellow colonies.

Propionibacterium acnes These anaerobic, Gram-positive rods are **diphtheroid,** or club shaped. When investigators tried to isolate an organism that might be the cause of acne, they almost always found the same Gram-positive diphtheroid rods in the lesions. Therefore, they named the organism *Propionibacterium acnes*. However, when people without acne were studied, it was found that *P. acnes* was present on their foreheads as well. Although some people have a much higher population of this organism than others, the elevated number does not seem to correlate with acne or any other skin condition.

Propionibacterium granulosum A Gram-positive diphtheroid rod found on some individuals, usually in smaller numbers than *P. acnes*. It is considered a harmless commensal.

Figure 22.1 Colonial appearance of some normal skin flora organisms in a Gram stain.

	Gram Stain	Colony Appearance
Staphylococcus Gram + coccus		White or gray (*epidermidis*) or Yellow (*aureus*)
Micrococcus		Yellow
Propionibacterium		White (*acne*) Buff (*granulosum*)

Definitions

Coagulase test. A test in which organisms are mixed with plasma on a slide. If the cells clump together, the culture is coagulase positive.

Commensals. Organisms that live together in close association and may or may not benefit each other.

Diphtheroid. A Gram-positive, club-shaped organism sometimes called a *coryneform*. *Propionibacterium* and *Corynebacterium* are examples of diphtheroid-shaped organisms.

Objectives

1. To learn to identify organisms making up the normal skin biota.
2. To understand the importance of skin biota.
3. To learn about the anaerobe jar.

Reference

Nester et al. *Microbiology: A human perspective,* 6th ed. The McGraw-Hill Companies, 2009. Chapter 23, Section 23.2.

Materials

First Session

TSY (trypticase soy yeast) agar plates (or TSY contact plates 2" in diameter), 2

GasPak anaerobe jar (or other anaerobic system)

Sterile swab

Sterile saline

70% ethanol

Cotton balls

Second Session

Tubes of sterile water (0.5 ml/tube), 6

TSY + glucose agar deeps (yeast extract and glucose are added to TS agar to encourage the growth of *Propionibacterium*), 4

TSY + Glucose + bromcresol purple agar slants, 2

Magnifying glass is optional but helpful

Third Session

Plasma

Procedure

Safety Precautions: Some students may isolate *Staphylococcus aureus* as part of their normal biota. This is a pathogen and should be handled with extra care.

First Session

1. Saturate a cotton ball with 70% ethanol and rub your forehead for 20 seconds. This will remove any transient organisms you might have on your skin, those that are not part of your normal flora.
2. Let your forehead dry for about 20 minutes. Avoid touching it with your hair or fingers.
3. Moisten a sterile swab with saline and rub it briskly on an area of your forehead about the size of a quarter for about 15 seconds.
4. Immediately swab the first third of a TSY agar plate, discard the swab, and finish the streak plate with a loop.
5. Repeat the procedure, swabbing a second TSY agar plate from an adjacent area of your forehead.

    (Alternative method to steps 3, 4, and 5: Press an open contact TSY agar plate on your forehead. Repeat with a second plate on an adjacent area.)
6. Incubate one of the TSY agar plates aerobically at 37°C. Incubate the second TSY agar plate anaerobically in a GasPak or other anaerobe jar at 37°C. Follow the manufacturer's directions for creating an anaerobic atmosphere.

7. After 48 hours of incubation, you or the instructor should store the aerobic plate at room temperature to prevent the plate from drying out. *Staphylococcus* and *Micrococcus* can be observed after 48 hours, but *Propionibacterium* must be incubated 5 days before colonies can be seen.

Second Session
(5 days later)

1. Examine the aerobic TSY plate and circle two different colony types with a marking pen on the bottom of the plate. Make a Gram stain of part of each circled colony.
2. If the colonies are Gram-positive cocci, suspend the remainder of the colony used for the Gram stain in 0.5 ml sterile saline. Use this suspension to inoculate:
   a. a glucose + bromcresol purple TS agar slant
   b. a cooled melted agar deep (sometimes called a shake tube). See exercise 9 for the inoculation procedure. Incubate at 37°C.
3. Examine the plate incubated in the anaerobe jar. You will see some of the same colony types observed on the aerobic plate because *Staphylococcus* are facultative anaerobes and can grow with or without oxygen. *Propionibacterium acnes* colonies, however, are white and very small—only a few millimeters in diameter. *Propionibacterium granulosum* are slightly larger and appear as buff or pink cones when the plate is examined with a magnifying glass. Choose two possible *Propionibacterium* colonies, Gram stain them, and if they are diphtheroid Gram-positive rods, inoculate into a shake tube. You can inoculate the shake tube directly from the plate if very little remains of the colony.

Third Session
(5 days later)

1. Observe the glucose + bromcresol slants. If the organism is able to ferment glucose, the acid produced will turn the purple agar yellow.
2. Observe the agar deeps. Obligate aerobes are only able to grow on the top, while facultative anaerobes can grow throughout the entire tube. The obligate anaerobes cannot grow in the top few centimeters where oxygen has diffused—only in the bottom anaerobic portion (see exercise 9).
3. Record your results and identify your isolates. The following is a description of the organisms most commonly found on the forehead.

Staphylococcus are Gram-positive cocci arranged in clusters. They are facultative and can ferment glucose. There are two main species of *Staphylococcus* found on the skin: *epidermidis* and *aureus*. *S. aureus* tends to have yellow colonies, and *S. epidermidis* white colonies.

Optional: If you have yellow colonies of *Staphylococcus*, you can determine if the species is *S. aureus* with a coagulase test. *S. aureus* is coagulase positive, and *S. epidermidis* is coagulase negative.

Micrococcus are Gram-positive cocci arranged in packets of four or eight. They are obligate aerobes and cannot ferment glucose or other sugars. They form round yellow colonies.

Propionibacterium acnes are diphtheroid coryneform Gram-positive rods that tend to palisade, or line up like a picket fence. They form very small, white, glistening colonies. *Propionibacterium* are obligate anaerobes that grow only in the absence of oxygen, although they are not killed by air as are some anaerobes.

Propionibacterium granulosum appears the same as *P. acnes* in a Gram stain but forms pinkish, slightly larger colonies.

Coagulase Test

Place a drop of water on a slide and make a *very* thick suspension of cells from a yellow colony.

Place a drop of plasma next to it, and mix the two drops together. Look for clumping; clumped cells indicate a coagulase-positive result.

Drop the slide in boiling water, and boil for a few minutes to kill the organisms before cleaning the slide.

See Appendix 4 for an alternative method.

	Summary of Reactions			
	Gram Stain	Colony Color	Metabolism	Glucose
Staphylococcus epidermidis	+ cocci	white/gray	facultative	+
Staphylococcus aureus	+ cocci	yellow	facultative	+
Micrococcus	+ cocci	yellow	obligate aerobe	–
Propionibacterium	+ rods	white/buff	obligate anaerobe	

EXERCISE

22

Laboratory Report: Normal Skin Biota

Results

	Isolate 1	Isolate 2	Isolate 3	Isolate 4
Gram stain				
Colony appearance				
Glucose fermentation				
Agar deep				
Possible identity				

Although this was not a quantitative procedure, what organism seemed to be the most numerous on your forehead?

Questions

1. How could normal skin biota be helpful to the host?

2. Why did you wipe your forehead with ethanol before sampling it?

3. Why was *Staphylococcus* the only organism that could grow on both plates?

4. How can you immediately distinguish *Staphylococcus* from *Propionibacterium* in a Gram stain?

5. Why does *Staphylococcus* probably cause more contamination than *Propionibacterium*, even though most people have higher numbers of the latter? (Hint: Are most agar plates incubated aerobically or anaerobically?)

EXERCISE

23

Respiratory Microorganisms

Getting Started

In this exercise, you will have an opportunity to observe a throat culture and learn some of the ways a clinical microbiologist identifies pathogenic organisms. You will also examine the normal biota of the throat, and while you will not work with actual pathogens, many of the organisms in the normal biota resemble related pathogenic bacteria. In exercises 28 and 29, you will use serological techniques to further differentiate some of these organisms.

A physician frequently orders a throat culture if a patient has a very sore throat and fever. This is done to determine whether or not the sore throat is caused by Group A β-hemolytic *Streptococcus pyogenes*. This organism is important because it not only causes strep throat, a disease characterized by fever and sore throat, but also two very serious diseases can follow the original infection: rheumatic fever, a disease of the heart, and acute glomerulonephritis, a disease of the kidney.

Fortunately, streptococci are usually still sensitive to penicillin and related antibiotics, and treatment is fairly straightforward. However, most sore throats are caused by viruses. Since viruses do not have a cell wall or any metabolic machinery, they are not affected by penicillin or most other antibiotics. Therefore, it is important to make an accurate diagnosis so that antibiotics can be used wisely.

Other pathogens sometimes found in the throat are *Streptococcus pneumoniae*, *Neisseria meningitidis*, *Haemophilus influenzae*, and *Klebsiella pneumoniae*. If these organisms are indeed causing disease, they are usually present in large numbers and the patient has the symptoms of the disease.

β-hemolytic *Streptococcus* can be identified when growing on blood agar. This medium is made of a base agar that is rich in vitamins and nutrients. Before pouring the melted agar into the petri plates, 5% sheep blood is added. (Blood is supplied from sheep kept for this purpose.) The blood performs two functions: it adds additional nutrients, and it aids in distinguishing α-hemolytic from β-hemolytic streptococci.

Streptococci produce hemolysins, which act on red blood cells (also called erythrocytes). α-hemolytic streptococci, which are a major component of the normal throat flora, incompletely lyse red blood cells. When the organism is growing on blood agar, a zone of partial clearing of the red blood cells can be seen around a colony (see color plate 6*a*). β-hemolytic streptococci produce hemolysins that completely lyse the red blood cells and therefore produce a clear zone in the blood agar around the colony (see color plate 6*b*). It is important to note that β-hemolysis is not always correlated with pathogenicity. For example, some strains of *E. coli* can produce β-hemolysis but are not responsible for any disease.

Commercial kits based on specific antibodies or other methods are now available for use in determining if a patient has strep throat. Although these tests can be performed in the doctor's office, the results sometimes must be verified by growing the culture on blood agar.

A small percentage of the hemolysins of β-hemolytic streptococci are oxygen labile, which means they are destroyed by oxygen. In a clinical laboratory, throat cultures are incubated in an anaerobe jar so that no hemolysis is overlooked. In this laboratory exercise, we will incubate the cultures in a candle jar, if convenient. This increases the CO_2 content of the atmosphere and enhances the growth of some organisms but does not create anaerobic conditions.

The throat contains a plethora of organisms that make up the normal biota, many of which resemble pathogenic organisms. Frequently, some actual pathogenic organisms are found in small numbers, such as *Streptococcus pneumoniae* or β-hemolytic *Streptococcus*. The presence of these organisms is only significant when they appear in large numbers.

Figure 23.1   Normal biota of the throat.

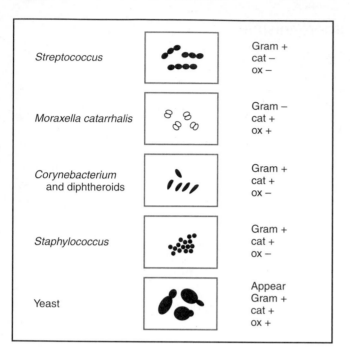

Streptococcus		Gram + cat − ox −
Moraxella catarrhalis		Gram − cat + ox +
Corynebacterium and diphtheroids		Gram + cat + ox −
Staphylococcus		Gram + cat + ox −
Yeast		Appear Gram + cat + ox +

Examples of common normal biota of the throat are diagrammed in figure 23.1. You might observe some of these on your streak plate:

α-hemolytic streptococcus   These organisms will be predominant on your plates. Another name for them is Viridans streptococci, and they rarely cause disease. They are Gram-positive and grow in short chains. They are catalase negative, which distinguishes them from *Staphylococcus*. *Streptococcus pneumoniae* is also *α*-hemolytic and is differentiated from the normal biota by sensitivity to certain antibiotics and serological tests. They are all **oxidase** negative.

Moraxella catarrhalis   These organisms are Gram-negative cocci arranged in pairs. They resemble the microscopic morphology of *Neisseria* (which grows only on a different kind of heated blood agar called chocolate agar). The genus *Neisseria* includes such pathogens as *N. gonorrhoeae* and *N. meningitis*. Try to find a colony of *Moraxella* and Gram stain it so you become familiar with the appearance of these organisms. The colonies may be a little larger than *Streptococcus* and are oxidase positive, so if oxidase reagent is available, it will help you identify a colony. They are also **catalase** positive.

Note: *Moraxella catarrhalis* was formerly named *Branhamella catarrhalis*, and prior to that, *Neisseria catarrhalis*.

Corynebacterium and Diphtheroids   These are irregular, club-shaped Gram-positive rods. They are part of the normal biota but resemble *Corynebacterium diphtheriae*, which causes diphtheria. They are usually catalase positive and oxidase negative.

Staphylococcus   These Gram-positive cocci are arranged in clusters. *Staphylococcus aureus* frequently is part of the normal biota, although it is a potential pathogen. The colonies are usually yellow. *Staphylococcus* is catalase positive and oxidase negative.

Yeasts   These are fairly common in the oral biota, forming relatively large colonies. In a Gram stain, the cells (which are eukaryotic) appear purple, are larger than bacteria, and sometimes have buds. They are catalase positive and usually oxidase positive.

Definitions

Catalase.   An enzyme found in most aerobic organisms that breaks down H_2O_2 to water and oxygen.

Oxidase.   A reagent that tests for cytochrome C.

Serological test.   Identification of organisms by mixing cells with serum containing antibodies to a specific organism. If the cells clump, the test is positive for that organism.

Objectives

1. To learn the importance of Group A *β*-hemolytic *Streptococcus* and how to distinguish it from normal biota.
2. To observe normal biota of the throat.

References

Holt, John G. et al. *Bergey's manual of determinative bacteriology.* Baltimore: Williams & Wilkins, 1994.

Nester et al. Microbiology: A human perspective, 6th ed. The McGraw-Hill Companies, 2009. Chapter 22.

Ryan, Kenneth J., ed. *Sherris medical microbiology, an introduction to infectious diseases,* 4th ed. 2003. McGraw-Hill.

Materials

Per student

First Session

    Blood agar plate, 1

    Sterile swab, 1

    Tube of sterile saline, 1

    Tongue depressor

    Demonstration cultures of the following:

        α-hemolytic *Streptococcus* on blood agar

        β-hemolytic *Streptococcus* on blood agar

    Candle jar

Second Session

    H_2O_2

    Oxidase reagent

Procedure

Safety Precautions: There may be colonies of β-hemolytic streptococci (*Streptococcus pyogenes*) and *Staphylococcus aureus* on the agar plates of normal biota. Handle these plates and the demonstration plates of β-hemolytic *streptococcus* with special care.

First Session

1. Swab your partner's throat. First seat your partner on a stool and shine a light into his or her throat. Carefully remove a sterile swab from the wrapper and moisten with the sterile saline, pressing out excess moisture on the inside of the tube. Depress the tongue with the tongue depressor and swab the tonsilar area on the side of the throat (figure 23.2). Do not swab the hard palate directly in the back behind the uvula and do not touch the tongue or lips. Do this rather quickly to avoid the gag response.

2. Swab the first third of a streak plate on the blood agar plate, rolling the swab over the agar to be sure to inoculate all sides. Discard the swab and continue streaking the rest of the plate with your loop as usual for isolated colonies.

3. Incubate your plate at 37°C in a candle jar, if available.

Figure 23.2 Diagram of open mouth. Shaded areas indicate places to swab.

Second Session

1. Examine your throat culture plate and compare it to the pure cultures of α- and β-hemolytic streptococci on the demonstration plates. If you observe β-hemolysis, make a Gram stain of part of the colony to determine if it is a Gram-positive streptococcus. If they indeed appear to be β-hemolytic streptococcus colonies, are they the predominant colony type on the plate? In a clinical laboratory:

    a. If they were the most numerous colony type, the physician would be notified and the patient would be treated.

    b. If only a few colonies were present, then the results could be reported as "ruled out β strep."

2. Make Gram stains of various colonies and do catalase and oxidase tests on the same colonies if possible. Record the kinds of organisms you observed based on their Gram stain, colony morphology, catalase test, and oxidase test. See Getting Started for a description of some of the organisms you might see.

Tests

Catalase. With a sterile loop, place some cells from the colony to be tested on a glass slide. Cover the cells with a few drops of H_2O_2. If bubbles form, the culture is catalase positive (figure 23.3a). Boil the slide in water for a few minutes to kill the organisms.

Oxidase. Place a small piece of filter paper on a glass slide and moisten with freshly prepared oxidase reagent. Remove some cells from a colony to be tested with a sterile loop (or sterile wooden stick) and rub the loop (or stick) on the moistened filter paper. If a pinkish-purple color appears, the cells are oxidase positive (see figure 23.3b). Place the paper in an autoclave bag or boil the slide for a few minutes to kill the organisms.

Figure 23.3 (a) Catalase and (b) oxidase tests.

(a)   Bubbles of O_2 in H_2O_2 indicate a positive catalase test.

(b)   Cells rubbed on paper turn dark pink indicating a positive oxidase test.

EXERCISE

23

Laboratory Report: Respiratory Microorganisms

Results

Indicate the numbers of colonies, using +++ for the most numerous or highest number, and + if very few are observed. Use ++ for numbers in between.

	Appearance of Hemolysis	Numbers Present
α-hemolytic streptococci		
β-hemolytic streptococci		

Other organisms observed:

Possible Identity	Gram Stain	Catalase	Oxidase	Numbers Present

Questions

1. What is the predominant organism in your throat flora? Did you observe any β-hemolytic streptococci in your throat culture?

2. What is the difference between α- and β-hemolysis
   a. on the red blood cells?

   b. on the blood agar plate?

3. Give two reasons it is very important to correctly diagnose and treat strep throat.

4. Name one genus of Gram-negative cocci.

5. If a student had a cold and sore throat caused by a virus, how would the virus appear on the blood agar plate?

EXERCISE

24

Identification of Enteric Gram-Negative Rods

Getting Started

In this exercise, you will learn to identify bacteria using biochemical tests. Gram-negative rods are excellent organisms to use as examples of this procedure because they are frequently seen in the clinical laboratory, both as pathogens or contamination from the normal biota of the large intestine.

The organisms in the colon are present in enormous numbers (about 10^{11}/gram of fecal material), but in healthy individuals these bacteria are part of the normal biota. Their presence prevents other organisms from becoming established and causing disease. An example of this situation occurs when the normal biota is eliminated after the therapeutic use of antibiotics. This permits *Clostridium difficile* to colonize the colon and cause life-threatening diarrhea.

Sometimes the normal biota is not so helpful. If certain of these organisms are introduced into the blood or infect the bladder or wounds, they can cause serious disease, especially in immunocompromised patients. These organisms, which are nonpathogens under usual circumstances but cause disease in special situations, are called **opportunists.**

Other **enteric** organisms are pathogens, such as *Salmonella,* the cause of diarrhea and typhoid fever, and *Shigella,* the cause of dysentery. It is important to identify these in a clinical laboratory, but they are not included in this exercise because of their pathogenicity.

The following list is a description of the enteric organisms used in this exercise:

Escherichia coli A small but constant part of the intestinal flora of almost all humans. It is usually nonpathogenic, but as an opportunist it is the major source of urinary tract infections from the patient's own biota. Many strains can cause diarrhea, especially in children and travelers. The notorious strain 0157:H7 causes bloody diarrhea and kidney failure. However, most strains are harmless and easily grown, making it one of the major microorganisms used in research.

Enterobacter Found in the colon, but also isolated from plant material and other sources in the environment.

Klebsiella Part of the normal flora of the colon, but some strains can cause pneumonia. It is also the cause of a small percentage of urinary tract infections.

Proteus Nonpathogen found in the colon, but the cause of a small percentage of urinary tract infections.

Pseudomonas A Gram-negative rod found in the normal biota of the colon, as well as in soil. It can cause serious wound infections, especially in burn patients. It is especially resistant to antibiotics.

All of these organisms, except *Pseudomonas,* belong to the family *Enterobacteriaciae,* defined as Gram-negative rods that are facultative anaerobes and able to ferment sugars. *Pseudomonas* is an obligate aerobe and cannot ferment sugars.

It is interesting to note that the major component of the colon biota is *Bacteriodes,* an obligate anaerobic Gram-negative rod. These organisms are rarely seen, however, because most laboratory procedures involving enteric bacteria do not include incubating plates anaerobically.

Identification Procedures

Biochemical tests measure such things as the differences in the organisms' ability to ferment different sugars, the presence of various enzymes, and physical characteristics such as motility. An organism then can be identified by comparing the results of the tests to the results of known bacteria in such reference books as *Bergey's Manual of Systematic Bacteriology.* The following is a discussion of the tests you will be using and how they work.

Fermentation Tubes Carbohydrate fermentation tubes consist of a complete broth; a carbohydrate, such as the sugar glucose; and the pH indicator andrades. A small Durham tube is added; fitting inside the large tube. If the organism can ferment the sugar, it will produce acidic products and the

Figure 24.1 Fermentation pathways.

Mixed Acid Fermentation

Glucose ⟹ Pyruvate

Pyruvate → Acetyl-CoA

Acetyl-CoA → Ethanol + Acetate

Acetyl-CoA → Formate

Formate → CO₂ + H₂

End products (relative molar amounts)

Acidic	4
Neutral	1
CO₂	1
H₂	1

Butanediol Fermentation

Glucose ⟹ Pyruvate

Pyruvate → Acetoin → 2,3-Butanediol + CO₂

Pyruvate → Ethanol

Pyruvate → Lactate

Pyruvate → Succinate

Pyruvate → Acetate

Pyruvate → CO₂ + H₂

End products (relative molar amounts)

Acidic	1
Neutral	6
CO₂	5
H₂	1

pH will fall, turning the pH indicator pink. If the organism also produces a gas, such as hydrogen, some of it will accumulate in the Durham tube as a bubble (see color plate 11).

After incubation, you should examine the tubes for growth (sometimes you have to shake the tubes slightly because the cells have settled). If for some reason there is no growth, the test must be repeated. If there is growth, the results can be recorded as:

A   Acid production—if the indicator has turned red

A/G   Acid and Gas—if the indicator has turned red and a bubble is seen in the Durham tube

N/C   No Change—if neither gas nor acid has formed

The sugars glucose, sucrose, and lactose are especially useful in the identification of the enteric Gram-negative rods. Lactose fermentation aids in the preliminary differentiation of enteric pathogens from the normal **coliforms.** *Salmonella* and *Shigella* species, which cause enteric diseases of various kinds, do not ferment lactose; however, members of the common fecal biota, *Escherichia coli, Enterobacter aerogenes,* and *Klebsiella,* are able to ferment lactose. Lactose fermentation is not related to pathogenicity in any way; it is simply a convenient characteristic for identifying organisms. The enteric *Proteus,* for example, is lactose negative but usually a nonpathogen.

Methyl Red and Voges-Proskauer   The fermentation tubes previously described show whether or not fermentation has taken place. Two tests identify a particular kind of fermentation: the methyl red test and the Voges-Proskauer test.

Fermentation is the energy-yielding pathway a facultative organism can use if oxygen is unavailable. Organisms have different fermentation pathways, resulting in different end products. Some are simple, such as the conversion of pyruvate to lactic acid (also termed lactate). Other pathways are more complex, yielding a variety of products and perhaps additional energy (figure 24.1).

The methyl red test measures the pH after organisms have grown in buffered peptone glucose broth (MR-VP broth). It is positive when the pH is less than 4.3. *E. coli* and other organisms ferment sugars by the mixed acid pathway. The products of this pathway are predominantly acetic and lactic acids, some organic compounds, and CO_2 and H_2. This results in a very low pH; therefore, *E. coli* is methyl red positive.

Other bacteria, such as *Enterobacter,* use the butanediol pathway when fermenting sugars. The end products are predominantly alcohols and a small amount of acid, as well as CO_2 and H_2. Although the pH is low, it is not low enough to change the methyl red, so these bacteria are methyl red negative.

One of the intermediates in the butanediol pathway is acetoin, for which the Voges-Proskauer reagents test. If the test is positive, then acetoin is present, and the organism is using the butanediol pathway. The reagents Voges-Proskauer A and B are added to the broth culture after incubation. If a brick-red precipitate forms, it is VP positive.

Indole   Some organisms have an enzyme that cleaves the amino acid tryptophan, producing indole. They can be grown in tryptone broth, which has a high level of tryptophan. After incubation, the broth is tested for the presence of indole by adding Kovacs reagent. A red ring forms on the top of the broth if the organism is indole positive. See color plate 12.

Citrate Utilization   The organism is grown on Simmons citrate medium, a mineral medium with citrate as the sole carbon source. The agar also contains the pH indicator bromthymol blue. If the organism can utilize the citrate, the pH rises and the indicator turns a deep blue. See color plate 13.

Urea Hydrolysis   Organisms are grown on agar containing urea and a pH indicator. If the organism produces the enzyme urease, urea is split, forming ammonia and CO_2. This raises the pH of the medium, turning it bright pink. *Proteus* species can be distinguished from other enteric rods because it always produces urease. See color plate 14.

Motility   Organisms are stabbed into a semisolid agar deep containing tetrazolium, an electron acceptor that turns red wherever there is growth. After incubation, a diffuse red color throughout the medium indicates motility. If there is a red streak only where the agar was stabbed, the organism is nonmotile.

MacConkey Agar   This is both a selective and differential medium. It is selective because only Gram-negative organisms can grow on it, and it is differential because colonies that ferment lactose are pink and non-lactose fermenting colonies are cream colored.

In this exercise, each team will inoculate a series of biochemical tests with a labeled organism and an unlabeled "unknown" organism. (The "unknown" is one of the organisms listed.) There may seem to be a large number of tubes to inoculate, but if they are labeled and lined up in a test tube rack, inoculation can be done fairly quickly. Good organization is very helpful.

Note: Commercial test systems, such as *Enterotube II System*, are available. A whole series of tests are inoculated at the same time with such systems (see color plate 15). After incubation, the results are read and the organism can be identified.

Definitions

Coliform.   Gram-negative rod found in the intestine that ferments lactose with the production of acid and gas—for example, *Escherichia coli*, *Enterobacter*, and *Klebsiella*.

Enteric.   Associated with the intestine (p. 199).

Fermentation.   An energy-yielding metabolic pathway in which organic compounds are both electron donors and acceptors.

Opportunist.   An organism that normally is nonpathogenic but can cause disease under special conditions.

Objectives

1. To learn to identify bacteria using biochemical tests.
2. To understand the physiological basis for the tests.
3. To become familiar with organisms commonly seen in a clinical laboratory, especially those enteric organisms causing urinary tract infections.

Reference

Nester et al. *Microbiology: A human perspective*, 6th ed. The McGraw-Hill Companies, 2009. Chapter 10, Section 10.2.

Materials

Per team of two students

Cultures on TS agar slants of the following:

Escherichia coli

Enterobacter aerogenes

Klebsiella pneumoniae

Pseudomonas aeruginosa

Proteus mirabilis

Sterile saline 1.0 ml/tube, 2 tubes

Glucose fermentation tubes, 2

Lactose fermentation tubes, 2

Sucrose fermentation tubes, 2

Tryptone broths, 2

Methyl red-Voges-Proskauer (MR-VP) broth tubes (one tube for each test), 4

Simmons citrate slants, 2

Motility agar deeps, 2

MacConkey agar plates, 2

Urea slants, 2

Reagents

    Oxidase reagent (freshly prepared) and small squares of white filter paper

    Kovacs reagent

    Methyl red

    Voges-Proskauer reagents A and B (V-P A and B)

Procedure

First Session

1. Choose one labeled culture and one "unknown" culture (or the cultures you are assigned).

2. Label a set of tubes for each organism with the name of the organism, the medium, the date, and your name or initials. Note that you will need 1 MR-VP broth for the methyl red test and another MR-VP broth for the Voges-Proskauer test. You should have a total of ten tubes for each organism to be identified.

3. Make a suspension of each organism by adding bacteria with a sterile loop to the sterile saline. Use this suspension to inoculate the test media.

4. Inoculate the tubes by adding a loopful of the suspended organisms to the broth. Inoculate the agar slant by gliding the loop over the surface of the agar, starting at the bottom of the slant. Streak the MacConkey plates with a loopful of suspension to obtain isolated colonies. Use good aseptic technique to avoid contamination with unwanted organisms.

5. Inoculate the motility deep by stabbing the agar almost to the bottom with your inoculating loop. Use cells from the original slant instead of from the suspension to be sure you have enough cells.

6. Perform an oxidase test using cells from the slant (see fig. 23.3, exercise 23).

7. Incubate the tubes at 37°C for 2 days or more. Incubate the motility deeps at room temperature. Some organisms are not motile at 37°C.

Second Session

1. Examine the fermentation tubes and record as "A" if the pH indicator has turned red from acid production, "G" for gas production, and "N/C" for no change if neither acid nor gas has formed. Be sure there is growth in the tube before recording (see plate 11).

2. Add a dropperful of Kovacs reagent to the tryptone broth and shake slightly. A red layer on the top of the broth indicates a positive test for indole (see plate 12).

3. Add a few drops of methyl red to one of the MR-VP broths. A red color indicates a positive test.

4. Add 10–15 drops of V-P reagent A (alpha-naphthol solution) and 10–15 drops of V-P reagent B (40% KOH). Shake and let stand a few minutes to an hour maximum. The appearance of a red color indicates a positive test.

    Safety Precaution: Alpha-naphthol is toxic.

5. Observe the citrate slant. A deep blue color is positive for citrate utilization (see plate 13).

6. Examine the motility deep. If the tube appears pink throughout the agar, it is positive for motility. If only the original stab line appears pink, the test is negative for motility.

7. Observe the colonies on the MacConkey plate. Pink colonies indicate lactose fermentation.

8. Observe the urea slant. A bright pink color is positive for urea hydrolysis (see plate 14).

9. Record the results of your known organism in the chart on the next page and on a similar chart drawn on the blackboard. These will be the results to establish the reactions of the known bacteria. Consult your instructor if student results of known bacteria do not agree.

10. Record the class results on the board. Also record the results of your unknown organism.

	Fermentation										
	glu	lac	suc	indole	MR	VP	cit	mot	ox	mac	urea
E. coli											
Enterobacter											
Klebsiella											
Proteus											
Pseudomonas											

NOTES:

EXERCISE

24

Laboratory Report:
Identification of Enteric Gram-Negative Rods

Results

	Fermentation			indole	MR	VP	cit	mot	ox	mac	urea
	glu	lac	suc								
E. coli											
Enterobacter											
Klebsiella											
Proteus											
Pseudomonas											
Unknown											

Questions

1. What is the identity of your unknown organism?

2. Can you determine whether or not an organism can ferment a sugar if it does not grow in the broth? Explain.

3. How can an organism have a positive test for acid from glucose in a fermentation tube, but have a negative methyl red test, which is also a glucose fermentation test?

4. Were there any organisms that did not ferment any sugars? If yes, which organisms?

5. When comparing a lactose fermentation tube with a MacConkey plate,

   a. what additional information does a fermentation tube give?

   b. what additional information does a MacConkey agar plate give?

EXERCISE

25

Clinical Unknown Identification

Getting Started

In this exercise, you will have an opportunity to utilize the knowledge and techniques you have learned in the previous three exercises to identify a mixture of two unknown organisms. You are given a simulated (imitation) **clinical specimen** containing two organisms, and your goal is to separate them into two pure cultures and identify them using different media and tests. The organisms are either associated with disease or are common contaminants found in normal biota or the environment.

Your unknown specimen represents either a urine infection or a wound infection. In actual clinical cases, standardized procedures exist for each kind of specimen. However, you will be identifying only a limited number of organisms. With some careful thought, you can plan logical steps to use in identifying your organisms.

The following are characteristics useful in identifying your unknown organism.

Bacterial Cell Morphology The size, shape, arrangement, and Gram-staining characteristics of the bacteria as determined by the Gram stain. It also may include the presence of special structures, such as endospores.

Colonial Morphology The appearance of isolated colonies on complete media, such as TS agar or blood agar, including their size, shape, and consistency.

Growth on Selective Media The ability of organisms to grow on **selective media.** Mannitol salt, which selects for organisms tolerating 7.5% salt; eosin methyleneblue (EMB), which selects for Gram-negative organisms; and MacConkey, which selects for Gram-negative rods, are examples of selective media.

*Reactions on **Differential Media*** The color of colonies on EMB agar or MacConkey agar is based on lactose fermentation (lactose fermenters are pink or purple). The appearance of organisms on mannitol salt agar is based on mannitol fermentation (mannitol fermenters turn the medium yellow). The

presence of hemolysis on blood agar constitutes another type of reaction of bacterial enzymes on red blood cells.

Biochemical Capabilities These include the ability to ferment different carbohydrates and the production of various end products, as well as the formation of indole from tryptophan. Tests include methyl red, Voges-Proskauer, citrate utilization, urease, catalase, oxidase, and coagulase.

Approach the identification of your "unknown" clinical specimen with the following steps:

1. Make a Gram stain of the specimen.
2. Streak the broth on the complete medium and the selective medium that seem appropriate.
3. After incubation, identify two different colony types and correlate with their Gram reaction and shape. Also correlate the growth and appearance of the colonies on selective media with each of the two organisms.
4. Restreak for isolation. It is useless to do any identification tests until you have pure cultures of the organisms.
5. After incubation, choose a well-isolated colony and inoculate a TS agar slant to be used as your stock culture. Prepare a stock culture for each organism.
6. Inoculate or perform various tests that seem appropriate. Keep careful records. Record your results on the worksheets as you observe them.
7. Identify your organisms from the test results.

Definitions

Clinical specimen. Cultures encountered in a medical laboratory.

Differential media. Media that permit the identification of organisms based on the appearance of their colonies.

Selective media. Media that permit only certain organisms to grow and that aid in isolating one type of organism in a mixture of organisms.

Objectives

1. To give you an opportunity to apply your knowledge to a microbiological problem.
2. To give you insight into the procedures used to isolate and identify clinical specimens.
3. To teach you to be aware of the presence of contaminants or nonpathogens in clinical specimens.

Note: See exercises 22 and 24 for more information on these organisms and tests.

Materials

First Session

Unknown mixture labeled with hypothetical source (for each student or team of two students)

    Blood agar plate or TS agar plate, 1 per student

    MacConkey agar plate (or EMB agar plate)

    Mannitol salt agar plate

Second and Third Sessions

    TS agar plates

    Nutrient agar slants

    Citrate agar slants

    Urea slant

    Glucose + bromcresol purple agar slants

    Fermentation broths of glucose, lactose, and sucrose

    MR-VP broth for the methyl red and Voges-Proskauer test

    Tryptone broth for the indole test

    Kovacs reagent

    Voges-Proskauer reagents A and B

    Methyl red for methyl red test

    Plasma for coagulase test

    Staining material for endospores and capsules

    H_2O_2

References

Difco manual of dehydrated culture media, 10th ed. Detroit: Difco Laboratories, 1984.

Holt, John G. et al. *Bergey's manual of determinative bacteriology*. Baltimore: Williams & Wilkins, 1994.

Murray, Patrick R. et al. *Manual of Clinical Microbiology*, 9th ed. Washington D.C.: ASM Press, 2007.

Nester et al. *Microbiology: A human perspective*, 6th ed. The McGraw-Hill Companies, 2009. Chapter 23 and Chapter 26.

Procedure

First Session

1. Make a Gram stain of the broth culture. Observe it carefully to see if you can see both organisms. You can save the slide and observe it again later if you have any doubts about it. You can also save the broth, but one organism may overgrow the other.
2. Inoculate a complete medium agar plate such as TS agar or blood agar, and appropriate selective and differential agar plates. Use MacConkey agar (if you suspect the possibility of a Gram-negative rod in a urine specimen) or a mannitol salt agar plate (if you suspect *Staphylococcus* in a wound infection). Streak the plates for isolated colonies.
3. Incubate at 37°C.

Second Session

1. Examine the streak plates after incubation and identify the two different colony types of your unknown organisms, either on the complete medium or the selective media, wherever you have well-isolated colonies. Gram stain each colony type (organisms usually stain better on nonselective media). Also identify each colony type on the selective and differential media so that you know which organisms grow on the various media. Record their appearance on the differential media as well. It is helpful to circle colonies that you Gram stain on the bottom of the petri plate with a marking pen.
2. Restreak each organism on a complete medium (instead of selective media) for

isolation. This technique ensures that all organisms will grow and that you will be able to see if you have a mixed culture. Do not discard the original streak plates of your isolates—store at room temperature. If at some point your isolate does not grow, you will be able to go back to the old plates and repeat the test.

Third Session

1. Observe the plates after incubation. If your organisms seem well isolated, inoculate each one on a TS agar slant to use as your stock culture. If you do not have well-isolated colonies, restreak them. It is essential that you have a pure culture. Possible steps in identifying Gram-positive cocci follow.

2. Look at the possible list of organisms and decide which ones you might have, based on the information you have found so far.

These are a few of the test results. Your instructor may provide more. Plan work carefully and do not waste media using tests that are not helpful. For example, a urea slant would not be useful for distinguishing between *Staphylococcus epidermidis* and *Staphylococcus aureus*:

Simulated wounds
  Staphylococcus epidermidis
  Staphylococcus aureus
  Micrococcus luteus
  Pseudomonas aeruginosa
Simulated urine infection
  Escherichia coli
  Enterobacter aerogenes
  Proteus
  Enterococcus faecalis
  (plus wound organisms)

A partial listing of the unknown organisms' possible identities follows. Others can be added.

Gram-Positive Cocci	
Staphylococcus aureus Found in either urine or wounds	Gram-positive cocci in bunches, catalase positive, yellow colony, ferments glucose (acid) and mannitol, coagulase positive, salt tolerant
Staphylococcus epidermidis Contaminant	Gram-positive cocci in bunches, catalase positive, ferments glucose (acid) but not mannitol, coagulase negative, salt tolerant
Micrococcus Contaminant	Gram-positive cocci in packets, catalase positive, yellow colony, does not ferment glucose or mannitol, coagulase negative, salt tolerant
Enterococcus faecalis Found in urine	Gram-positive cocci in chains, catalase negative, ferments glucose (acid), coagulase negative

Gram-Positive Rod	
Bacillus	Large, gram-positive rods, forms spores, catalase positive

Gram-Negative Rods	
Escherichia coli Urine	Glucose and lactose positive (acid and gas), indole positive, methyl red positive, Voges-Proskauer negative, citrate negative, urea negative, oxidase negative
Proteus Urine	Lactose negative, oxidase negative, urea positive, ferments glucose (acid), indole negative
Pseudomonas Urine and wounds	Lactose negative, glucose negative, urea negative, oxidase positive, indole negative
Enterobacter	Glucose and lactose positive (acid and gas), indole negative, methyl red negative, Voges-Proskauer positive, citrate positive, urea negative, oxidase negative

NOTES:

EXERCISE

25

Laboratory Report: Worksheet and Final Report:
Clinical Unknown Identification

Gram stain of original specimen: _____

_____

(describe cell shape, arrangement, and Gram reaction)

Gram stains of TS agar subcultures: _____

_____

(describe cell shape, arrangement, and Gram reaction)

Test	Organism #1	Organism #2
Colony description		
(TS agar or blood)		
Gram stain		
Colony appearance MacConkey (or EMB)		
Colony appearance mannitol salt agar		
Special stains		
capsule		
endospore		
Lactose fermentation		
Glucose fermentation		
Sucrose fermentation		
Mannitol fermentation		

Test	Organism #1	Organism #2
Indole production		
Methyl red		
Voges-Proskauer		
Citrate utilization		
Urea hydrolysis		
Motility		
Catalase test		
Coagulase		
Oxidase		

Consult the table compiled from the class results at the end of the second session in exercise 24 and figure 23.1 to help you identify your organisms.

Final Identification
Organism #1

Organism #2

Final Report

1. What is the identification of your organisms? Discuss the process of identification (reasons for choosing specific tests, any problems, and other comments).

   Organism #1:

   Organism #2:

NOTES:

I NTRODUCTION to Some Immunological Principles and Techniques

Immunology, the study of the body's immune response, is responsible for protecting the body against disease. It is often triggered when foreign substances or organisms invade the body. Examples include pathogenic microbes and the chemical compounds they produce, such as foreign materials called **antigens.** In diseases such as AIDS and cancer, the body's immune response is either seriously weakened or destroyed, whereas for milder diseases the immune response remains complete.

Different forms of the immune response include **phagocytic cells,** such as white blood cells (WBCs); **enzymes,** such as **lysozyme;** and **antibodies.** WBCs and enzymes are examples of natural immunity since they are already present in the body and need not be triggered by the antigen. In contrast, antibodies are an example of acquired immunity since their formation is triggered only in the presence of the antigen. For some people, the immune response can be triggered by the body's own proteins. This can result in the formation of **autoimmune diseases,** such as rheumatoid arthritis and glomerulonephritis.

Phagocytic cells and enzymes are also examples of nonspecific immunity since they can react with a variety of different foreign substances (for instance, phagocytic cells can engulf both inanimate and animate particles). Conversely, antibodies represent a form of specific immunity because they are produced in response to particular antigens (an antibody produced against *Salmonella* cell walls will not react with *Proteus* cell walls).

The exercises include examples of both innate immunity (exercises 26 and 27) and acquired immunity (exercises 28–30). In exercise 26, you will study human blood cells and learn how to determine which ones are phagocytic. In exercise 27, you will learn how to determine the bactericidal activity of the enzyme lysozyme, which occurs naturally in phagocytic white blood cells, saliva, nasal secretions, and tears. Lysozyme is able to digest the cell walls of many bacteria.

The adaptive immunity exercises demonstrate examples of antigen-antibody reactions, which vary from one another depending on the nature of the antigen. If the antigen is particulate (as in cells or insoluble substances such as cardiolipin) an agglutination, or clumping, reaction will occur that can be observed visually. For nonparticulate antigens, a precipitin reaction occurs that can often be seen with the naked eye, eliminating the need to perform additional visualization procedures (for example, in exercise 30 the use of an agar immunodiffusion for *Coccidioides* identification, in which precipitin lines form in the agar at the site of the antigen-antibody reaction).

Exercise 29 represents the current state-of-the-art techniques for demonstrating particulate antigen-antibody reactions. One such reaction, latex agglutination, is commonly used for *S. aureus* detection. The particulate antigen-antibody reaction, enzyme-linked immunosorbent assay (ELISA), is described in exercise 29, where you will use it for *Coccidioides* identification. It is also widely employed in other areas, such as plant and animal virus identification and detection of HIV antibodies.

Definitions

Antibody.   A protein produced by the body in response to a foreign substance (*e.g.,* an antigen); it reacts specifically with that substance.

Antigen.   Any cell particle or chemical that can cause production of specific antibodies and combine with those antibodies.

Autoimmune disease.   An immune reaction against one's own tissues.

Enzyme.   A protein that acts as a catalyst. A catalyst is a substance that speeds up the rate of a chemical reaction without being altered or depleted in the process.

Lysozyme. An enzyme that degrades the peptidoglycan layer of the bacterial cell wall.

Phagocytic cells. Cells that protect the host by ingesting and destroying foreign particles, such as microorganisms and viruses.

Some of our cells, although they are part and parcel of us, have not even fixed coherence within our 'rest'. Such cells are called 'free'....The cells of our blood are as free as fish in a stream. Some of them resemble in structure and ways so closely the little free swimming amoeba of the pond as to be called amoeboid. The pond amoeba crawls about, catches and digests particles picked up in the pond. So the amoeboid cells inhabiting my blood and lymph crawl about over and through the membranes limiting the fluid channels in the body. They catch and digest particles. Should I get a wound they contribute to its healing. They give it a chance to mend, by eating and digesting bacteria which poison it and by feeding on the dead cells which the wound injury has killed. They are themselves unit lives and yet in respect to my life as a whole, they are components in that corporate life.

Sherrington, *Man on His Nature*

EXERCISE

26

Differential White Blood Cell Stains

Getting Started

This exercise deals with the cellular forms of the immune system, specifically the white blood cells (WBCs). For the most part, they can be distinguished from one another using a blood smear stained with Wright's stain, a differential stain. This stain uses a combination of an acid stain, such as eosin, and a basic stain, such as methylene blue. They are contained in an alcoholic solvent (methyl alcohol), which fixes the stains to the cell constituents, particularly since the basophilic granules are known to be water soluble. With this stain, a blood smear shows a range in color from the bright red of acid material to the deep blue of basic cell material. In between are neutral materials that exhibit a lilac color. There are also other color combinations, depending upon the pH of the various cell constituents.

The two main groups of WBCs are the granulocytes (cytoplasm contains granules) and the agranulocytes (clear cytoplasm). The granulocytes are highly phagocytic and contain a complex, segmented nucleus. The agranulocytes are relatively inactive and have a simple nucleus, or kidney-shaped nucleus. Common cell types found in the granulocytes are **neutrophils, eosinophils,** and **basophils.** The basic agranulocyte cell types are the **lymphocytes** and **monocytes.** Another WBC type found is the **platelet** (very small, multinucleate, irregular pinched off parts of a **megakaryocyte**). Platelets aid in the prevention of bleeding. The appearance of these cell types in blood stained with a differential stain are illustrated in figure 26.1.

Differential blood stains are important in disease diagnosis because certain WBCs either increase or decrease in number depending on the disease. In making such judgments, it is important to know the appearance of normal blood (see color plate 16). The microscopic field shown in color plate 16 includes mostly RBCs with a few neutrophils, only one lymphocyte, and some platelets.

Figure 26.1 Blood cell types present in human peripheral blood. The granular leukocyte names find their origin from the color reaction produced by the granules after staining with acidic and basic components of the staining solution. Neutrophil = neutral-colored granules; basophil = basic color; and eosinophil = acid color.

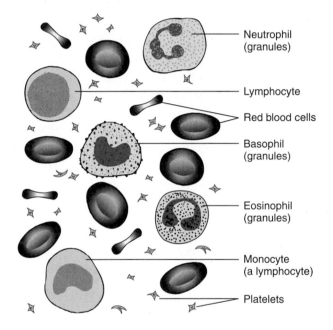

A quantitative description of the various cell types found in normal blood is shown in table 26.1. The red blood cells (RBCs), which are also called erythrocytes, make up the largest cell population. RBCs constitute an offensive weapon because they transport oxygen to various body parts and they

Table 26.1 Cellular Description of Normal Blood*

Total Counts	Differential WBC Counts
RBC 5,200,000/ml	Neutrophils 64%
WBC 7,200/ml	Leukocytes 33%
Platelets 350,000/ml	Monocytes 2%
Basophils 50/ml	Eosinophils 1%

*From R. Kracke. *Diseases of the blood and atlas of hematology,* Second Edition. Copyright 1941. J. B. Lippincott Co. By permission.

break down carbon dioxide to a less toxic form. The RBCs in humans and all other mammals (except members of the Family Camelidae, such as the camel) are biconcave, circular discs without nuclei.

RBCs are produced in the red bone marrow of certain bones. As they develop, they produce massive quantities of hemoglobin, the oxygen-transporting pigment that contains iron, which in the oxygenated form gives blood its red color. Worn-out RBCs are broken down at the rate of 2 million cells per second in the liver and spleen by phagocytic WBCs. Some of the components of the RBCs are then recycled in order for the body to maintain a constant number of these cells in the blood.

The WBCs, or leukocytes, represent approximately $\frac{1}{800}$ of the total blood cells (see table 26.1). They are defensive cells, specialized in defending the body against infection by microorganisms and other foreign invaders. Many of the leukocytes are **amoeboid,** capable of moving independently through the bloodstream. They also move out into the tissues, where they repel infection and remove damaged body cells from bruised or torn tissue.

In this exercise, you will have an opportunity to stain differentially and to observe some human blood slides. For this purpose, the blood used can be outdated whole blood obtained from a blood bank. If this is not available, commercially prepared blood slides are available.

Definitions

Amoeboid.   To make movements or changes in shape by means of protoplasmic flow.

Basophil.   A granulocyte in which the cytoplasmic granules stain dark purplish blue with methylene blue, a blue basophilic-type dye found in Wright's stain.

Eosinophil.   A granulocyte in which the cytoplasmic granules stain red with eosin, a red acidophilic-type dye found in Wright's stain.

Lymphocyte.   A colorless agranulocyte produced in lymphoid tissue. It has a single nucleus and very little cytoplasm.

Megakaryocyte.   A large cell with a lobulated nucleus that is found in bone marrow. It is the cell from which platelets originate.

Monocyte.   A large agranulocyte normally found in the lymph nodes, spleen, bone marrow, and loose connective tissue. It is phagocytic with sluggish movements. When stained with Wright's stain, it is difficult to differentiate from a junior neutrophil.

Neutrophil.   A mature granulocyte present in peripheral circulation. The cytoplasmic granules stain poorly or not at all with Wright's stain. The nuclei of most neutrophils are large, contain several lobes, and are described as polymorphonuclear (PMN) leukocytes.

Platelet.   A small oval-to-round, colorless, biconcave disc, 3 microns in diameter. Plays a role in clotting of blood.

Objectives

1. To provide historical and background information on blood and some of its microscopic cell types, their origin, morphology, number, and role in fighting disease.
2. To prepare two stained blood slides: the first for use in observing the cellular appearance of normal blood, and the second for use in determining a differential WBC count.

Reference

Nester et al. *Microbiology: A human perspective,* 6th ed. The McGraw-Hill Companies, 2009. Chapter 15, Introduction.

Materials

Either outdated blood bank whole blood or prepared commercial unstained or stained human peripheral blood smears. For student use, outdated whole blood should be dispensed with either a plastic dropper or a dropping bottle capable of dispensing a small drop.

Note: In the event of spilled blood, use disposable gloves and towels to remove blood.

Then disinfect the area with a germicide, such as hypochlorite bleach diluted approximately 1:20 with water.

For use with whole blood

New microscope slides, 3

Plastic droppers for dispensing blood on slides, dispensing Wright's stain, and adding phosphate buffer

Hazardous waste container for droppers and slides

For use with whole blood and unstained prepared slides

Wright's stain, dropping bottle, 1 per two students

Phosphate buffer, pH 6.8, dropping bottle, 1 per two students

Wash bottle containing distilled water, 1 per two students

A Coplin jar with 95% ethanol

Staining rack

Colored pencils for drawings: pink, blue, purple, or lavender

Procedure

Safety Precautions: Working with blood, even blood from the blood bank, carries risks from unknown contaminants. Disposable gloves and goggles must be worn when preparing slides.

1. Prepare three clean microscope slides free of oil and dust particles as follows:
   a. Wash slides with a detergent solution and then rinse thoroughly.
   b. Immerse slides in a jar of 95% alcohol.
   c. Air dry and polish with lens paper.
2. Place a drop of blood on one end of a clean slide (figure 26.2a). Repeat with a second clean slide.
3. Spread the drop of blood on the slide as follows:
   a. Place the slide on your laboratory bench top. With your thumb and middle finger, firmly hold the sides of the slide on the end where the drop of blood is located.
   b. With your other hand, place the narrow edge of a clean slide approximately ½" in *front* of the drop at an angle of about 30° (figure 26.2b).
   c. Carefully push the spreader slide *back* until it comes in contact with the drop, at which point the drop will spread outward to both edges of the slide (figure 26.2c).
   d. Immediately, and with a firm, steady movement, push the blood slowly toward the opposite end of the slide (figure 26.2d).

Note: Use of these procedural restraints (a small drop, a small spreader slide angle, and a *slow*, steady spreader slide movement) should provide a thin film for study of RBCs. A good smear has the following characteristics: smooth, without serrations; even edges; and uniform distribution over the middle two-thirds of the slide.

   e. Allow slide to air dry for 5 minutes. Do not blot.
   f. For the second slide, prepare a thicker film by using a larger spreader slide angle (45°) and by pushing the blood more rapidly to the opposite end of the slide. The second slide is best for determining the differential WBC count.

Note: The unused end of the first spreader slide can be used to prepare the second slide. Discard used spreader slide in the hazardous waste container.

4. Stain the blood smears with Wright's stain as follows:
   a. Suspend the slides so that they lie *flat* on the staining rack supports.
   b. Flood or add 15 drops of Wright's stain to each blood smear. Let it stand for 3–4 minutes. This fixes the blood film.
   c. Without removing the stain, add an equal volume of phosphate buffer. Blow gently through a pipet on each side of the slide to help ensure mixing of stain and buffer solutions.
   d. Let stand until a green, metallic scum forms on the surface of the slide (usually within 2–4 minutes).

Figure 26.2 Method for preparing a blood smear.

(a) A small drop of blood is placed about 3/4" away from one end of slide. The drop should not exceed 1/8" diameter.

(b) The spreader slide is moved in the direction of the arrow *(back)*.

(c) Drop of blood spreads outward to both edges of slide.

Thin smear

(d) The spreader slide is pushed along the slide, dragging the blood over its surface.

e. Wash off the stain with water. Begin washing while the stain is on the slide in order to prevent precipitation of scumlike precipitate, which cannot be removed. The initial purple appearance on the slide should be washed until it is lavender-pink.

f. Wipe off excess stain from the back of the slide and allow it to stand on end to dry (which is preferable to drying between bibulous paper).

5. Examine stained blood smears:
   a. Make an initial examination of the first blood smear with the low-power objective lens to find the most suitable areas for viewing with the oil immersion objective lens.
   b. Next, using the oil immersion lens, make a study of the various WBC types present: basophils, eosinophils, lymphocytes, monocytes, neutrophils, and platelets. For help in this study, consult color plate 16, the Definitions section describing their staining characteristics with Wright's stain, and figure 26.1. Prepare color drawings of your findings in the Laboratory Report.
   c. Conduct a differential WBC count using the second blood smear. For normal blood with a leukocyte count of 5,000–10,000 WBCs/ml, you would classify 100 leukocytes. In order to do this, you may have to examine the number and kinds of WBCs present in as many as twenty microscopic fields. Record your findings in table 26.2 of the Laboratory Report and calculate the percentage of each WBC type.

EXERCISE

26

Laboratory Report: Differential White Blood Cell Stains

Results

1. Color drawings of RBCs and various WBCs found in blood smears stained with Wright's stain.

   RBCs   Neutrophils   Eosinophils

   Basophils   Monocytes   Lymphocytes

2. Differential WBC count. In table 26.2, record the kinds of leukocytes found as you examine each microscopic field. After counting 100 WBCs, calculate their percentages from the totals found for each type. Also record the number of microscopic fields examined to find 100 WBCs: _____.

Table 26.2   Kinds and Percentages of WBCs Found in Blood Smear

Neutrophils	Eosinophils	Basophils	Lymphocytes	Monocytes
Total				
Percent				

Questions

1. What problems, if any, did you find in preparing and staining your blood smears? Indicate any differences noted between thin and thick smears.

2. Were your blood stains satisfactory? Did the stained cells resemble those in figure 26.1 and color plate 16? Were they better?

3. Did your differential white blood cell count percentages compare with the percentages in normal blood (table 26.1)? If not, give an explanation.

4. Were there any WBC types that you did not find in your blood smear? If so, which one(s)? Why did you not find them?

5. Matching (you may wish to consult your text).

   a. Neutrophils   ___ Involved in antibody production
   b. Basophils   ___ A minor phagocytic cell
   c. Monocytes   ___ Increased number in parasitic infections
   d. Eosinophils   ___ Largest WBC
   e. Lymphocytes   ___ Inflammatory WBC

EXERCISE 27

Lysozyme, an Enzyme Contributing to Innate Immunity

Getting Started

A number of antimicrobial substances have been isolated from animal cells and body fluids. One of these is the protein **lysozyme,** discovered in 1922 by Alexander Fleming. This is an important enzyme contributing to **innate immunity** in animals. Unlike adaptive immunity, which requires an animal to be first exposed to the pathogen for protection, innate immunity protects the animal at birth.

Lysozyme is a **proteolytic enzyme** found in tears, saliva, and nasal secretions. It is also found in phagocytic WBCs (see exercise 26) and in egg white. It is able to enzymatically degrade the **peptidoglycan layer** of the bacterial cell wall, thereby weakening and eventually destroying it. It is particularly active against Gram-positive bacteria that have an exposed peptidoglycan layer. Experimentally, when bacterial cells are treated with lysozyme in a solution of high osmotic pressure, such as 0.6 M sucrose, the cell wall dissolves, but cell lysis is prevented (see exercise 13).

Lysozyme is also found in the tail of bacteriophage, where it expedites entry of the phage into the cell. Later phage stages direct the host cell to produce more lysozyme for use in facilitating release of new phage particles.

In this exercise, you will have an opportunity to assay the antimicrobial activity of lysozyme collected from tears and egg white and to compare this activity with a commercial lysozyme preparation of known activity.

Definitions

Innate immunity. An immunity to infectious disease in a species occurring as a part of its natural biologic makeup.

Lysozyme. An enzyme able to attack and destroy bacterial cell walls. It occurs naturally in tears, saliva, phagocytic WBCs, and egg white.

Peptidoglycan layer. The rigid backbone of the bacterial cell wall, composed of repeating subunits of N-acetylmuramic acid and N-acetylglucosamine and other amino acids.

Proteolytic enzyme. An enzyme able to hydrolyze proteins or peptides with the resulting formation of simpler, more soluble products such as amino acids.

Objectives

1. To become more familiar with the process of innate immunity and how such chemicals as lysozyme function in immunity.
2. To learn how to determine the antimicrobial activity of various natural lysozyme preparations.

References

Fleming, A. 1922. Proc. of the Royal Soc. of London, Ser. B, 93, 306.

Nester et al. *Microbiology: A human perspective,* 6th ed. The McGraw-Hill Companies, 2009. Chapter 3, Section 3.6.

Osserman, et al., eds. *Lysozyme.* New York: Academic Press, 1974.

Materials

Per pair of students

Nutrient broth cultures (37°C, 24 hour) of *Staphylococcus epidermis* and *E. coli*

Sabouraud's dextrose broth culture (25°C, 24 hour) of *Saccharomyces cerevisiae*

Melted nutrient agar deeps, 2, and Sabouraud's dextrose agar deep, 1, held in a 48°C water bath

Sterile petri dishes, 4

Petri dish containing 9 sterile filter paper discs (approximately ½" diameter)

Petri dish containing 1–2 ml of aseptically prepared raw egg white

Petri dish containing 1–2 ml of lysozyme chloride (Sigma cat. # L-2879) with activity of approximately 60,000 units per mg of protein; diluted 1:10 with sterile distilled water

Test tube containing 9 ml sterile distilled water

Raw onion

Scalpel or sharp knife for cutting the onion

Mortar and pestle

Pair of tweezers

Sterile 1-ml pipets, 4

Ruler divided in mm

Procedure

Note: An alternative method is the use of confluent growth plates (see exercise 14) instead of pour plates.

First Session

1. With a glass-marking pen, divide the underside of three petri dishes in thirds. Label one part "T" for tears, another part "EW" for egg white, and the remaining part "L" for lysozyme.
2. Label each petri dish with the name of one of the three test organisms. Also include your name and the date on the bottom.
3. Prepare pour plates of *Staphylococcus* and *E. coli* by suspending the broth culture. With a sterile 1-ml pipet, aseptically transfer a 1-ml aliquot to the respectively marked petri dish. Discard the pipet in the hazardous waste container. Add the contents of a tube of melted nutrient agar to the dish, replace cover, and rotate the dish six to eight times on the surface of your desk to distribute the cells uniformly. Allow agar to harden. Repeat this procedure for the other culture.
4. Repeat step 3 above for *Saccharomyces cerevisiae*, using a melted Sabouraud's agar deep.
5. In order to induce tears, one student should remove the outer skin from an onion. Then cut the onion into small pieces and crush in a mortar with a pestle. The other student should be prepared to collect the secreted tears in a sterile petri dish (0.5–1 ml is sufficient).
6. Alcohol sterilize tweezers and air cool. Aseptically remove a filter paper disc and dip it into the tears. Transfer the moistened disc to the center of the area marked "T" on the *S. epidermis* agar plate.
7. Repeat step 6 by transferring moistened tear discs to petri dishes containing the remaining two test organisms.
8. Repeat steps 6 and 7 with the egg white preparation.
9. Repeat steps 6 and 7 with the lysozyme preparation.
10. Incubate the petri dishes containing *Staphylococcus* and *E. coli* at 37°C for 24 hours and the petri dish containing *Saccharomyces cerevisiae* at 25°C for 48 hours.

Second Session

1. Observe the petri dishes for zones of inhibition around the filter paper discs. With a ruler, determine their diameter in mm, and record your results in table 27.1 of the Laboratory Report.

EXERCISE

27

Laboratory Report:
Lysozyme, an Enzyme Contributing to Innate Immunity

Results

1. Complete table 27.1 (see instructions in Procedure, session 2, step 1).

Table 27.1 Antimicrobial Activity of Various Lysozyme Extracts

Test Organism	Diameter of Inhibition Zone (mm)		
	Tears	Egg White	Lysozyme*
S. epidermis			
E. coli			
S. cerevisiae			

*1:10 dilution

2. Which of the three preparations was the most active? _____ least active? _____ Consider the lysozyme *dilution* when preparing your answer.

3. Which organism(s) were not inhibited by lysozyme?

4. Knowing that this lysozyme preparation has an activity of 60,000 units per mg of protein, calculate the lysozyme activity in similar units for tears and egg white against *S. epidermis* and *E. coli*. Some of your results from table 27.1 are necessary for making this calculation. Record your findings in table 27.2.

Table 27.2 Units of Lysozyme Activity for Egg White and Tears

Test Organism	Units of Lysozyme Activity/Mg of Protein	
	Egg White	Tears
S. epidermis		
E. coli		

Questions

1. What similarity do lysozyme and penicillin have in their mode of antimicrobial action? How do they differ?

2. Why must tweezers be cool before dipping the filter paper discs in the lysozyme solutions?

3. What large groups of microorganisms are susceptible to lysozyme? resistant to lysozyme?

4. Procedurally, what additional important step would be necessary to evaluate the lysozyme activity of nasal secretions? Why?

5. Why are most Gram-negative bacteria not lysed by lysozyme, yet they have a peptidoglycan cell wall structure similar to that of the Gram-positive bacteria?

NOTES:

EXERCISE

Lancefield Grouping of Pathogenic Streptococci

Getting Started

The genus *Streptococcus* deserves special attention because of its involvement in numerous diseases of humans and animals (see exercises 22 and 23). It represents the predominant normal bacterial flora of the human respiratory tract, and it is also found in the intestinal and genital tracts. Few other microorganisms of medical importance can elaborate as many **exotoxins** and destructive enzymes, as well as produce serious infections in virtually every tissue, as *streptococci* (table 28.1). One of the **sequela** diseases, rheumatic fever, is caused by a Lancefield Group A streptococcus that initially causes pharyngitis (strep throat).

It is important to differentially diagnose strep throat infections from viral throat infections since both acute rheumatic fever and acute **glomerulonephritis** can occur if prompt appropriate therapy is not initiated. Rapid immunological slide tests are now available for detecting and differentiating pathogenic streptococci. They have their origin in historical research done by Rebecca Lancefield (1933), who was able to group streptococci serologically into eighteen groups (A–R) on the basis of a group-specific carbohydrate antigen present in their cell wall. One exception is the Group D streptococci, in which a noncarbohydrate antigen, teichoic acid, is found in the cytoplasmic membrane. The serological test developed by Lancefield was an overnight capillary **precipitin reaction** test using extracted cell wall antigens.

Further examination of table 28.1 shows that most human streptococcal diseases are caused by beta (β)-hemolytic streptococci (see exercise 23 for information on β-hemolysis determination) belonging to Lancefield Group A. This species, *S. pyogenes*, in contrast to other pathogenic streptococci, is found in only a small percentage of healthy humans. Other characteristics important for final identification of streptococci include biochemical and morphological tests (see exercises 22 and 23). In this exercise, you will examine streptococci morphology and hemolysis reactions on blood agar. With one exception—*S. pneumoniae*—all streptococci appear as chains when grown in a broth medium (figure 28.1a). On an agar medium, they sometimes appear as diplococci (figure 28.1b). *Streptococcus pneumoniae* is a true diplococcus that can be differentiated from other diplococci by its lancet shape (figure 28.1c) and **encapsulation.**

Table 28.1   Some Characteristics Useful in Identifying the Major Streptococci Associated with Human Disease

Streptococcus Species	Lancefield Group	Hemolysis Type	Human Diseases
S. pyogenes	A	β	Pharyngitis, scarlet fever, wound infections, skin, ear, lungs, other tissues, sequela: glomerulonephritis, rheumatic fever
S. agalactiae	B	β	During birth and in infants: may develop pneumonia, septicemia, meningitis
S. pneumoniae		α	Bacterial pneumonia, ear infections, meningitis
S. faecalis *S. faecium*	D	α or β	Subacute bacterial endocarditis (rare), urinary tract infections
S. durans, S. bovis	D	α or γ	Endocarditis, wound infections
S. viridans (10 species)	Occasional false +	α	Subacute bacterial endocarditis

(a)

(b)

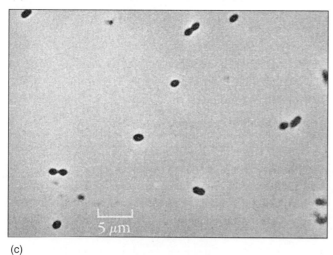

(c)

Figure 28.1 Genus *Streptococcus* morphology:
(*a*) *S. pyogenes* chains in a broth medium;
(*b*) *S. pyogenes,* sometimes diplococci on solid media;
(*c*) *S. pneumoniae,* lancet-shaped diplococci in a broth medium. Phase-contrast photomicrographs courtesy of G. E. Roberts.

Streptococci from the oral cavity can be semi-quantitated using a dental caries test. However, since they are normally considered to be opportunistic, in a clinical setting the species is not differentiated. The streptococci responsible for the majority of human diseases are *Streptococcus pyogenes, Streptococcus agalactiae, Streptococcus pneumoniae,* and *Enterococcus faecalis.* Differential tests are important tools for proper diagnosis and treatment of infection.

The principle of the Synder dental caries test is that the higher quantities of cavity-causing bacteria, like *Streptococcus mitis, Streptococcus mutans,* and *Streptococcus salivarius,* an individual possesses in the oral cavity, the sooner visualization of fermentation can be seen in the agar. Snyder agar contains bromocresol green as a pH indicator, so when acid is produced during fermentation, the dark blue medium changes to green, and finally to a yellow color (see color plate 27). Thus, the inoculated tubes will need to be inspected at 24, 48, and 72 hours. An individual tube that has turned yellow at 24 hours most likely harbors more cavity-causing streptococci than a tube that takes 72 hours to turn completely yellow, hence being semi-quantitated.

In order to differentiate streptococci presenting similar hemolysis results, several tests can be utilized. The bile esculin test is used to help differentiate Lancefield Group D (*Enterococcus faecalis*) from other Lancefield groups (*Streptococcus pyogenes*), although some of Group B test positive for this test. Certain genera of Enterobacteriaceae, such as *Serratia, Klebsiella,* and *Enterobacter,* can also test positive, so a gram stain is needed to ensure which is present—streptococci or Enterobacteriaceae. The bile salts make this media selective, and the hydrolysis of the esculin makes it differential. Thus, a color change from a light tan to dark brown indicates a positive reaction (see color plate 28). In conjunction with the bile esculin test, we must further differentiate the enterococci (*Enterococcus faecalis*) from nonenterococci (*S. bovis*) by using a Tryptic Soy broth tube containing 6.5% sodium chloride. *Enterococcus faecalis* grows in the high salt concentration, whereas *S. bovis* does not.

In order to differentiate *S. agalactiae* from other β-hemolytic streptococci (like *S. pyogenes*), a CAMP test can be performed. The CAMP test was named after the individuals (Christie, Atkins, and

Munch-Petersen) who first utilized the test. The principle of this test is that in the presence of staphylococci, more CAMP factor will be produced by *S. agalactiae* and an increased production of enzyme will occur, causing an arrowhead shape to be seen at the intersection of the two bacterial streaks.

To further differentiate Lancefield groups of streptococci from one another, antibiotic disks can be utilized. *Streptococcus pyogenes* is susceptible to bacitracin, whereas the other β-hemolytic streptococci, like *S. agalactiae*, are resistant. However, *S. pneumoniae* is normally considered susceptible to bacitracin, but it is α-hemolytic. In turn, *Streptococcus pneumoniae* is susceptible to the antibiotic optochin, but other α-hemolytic streptococci are resistant to it. Thus, antibiotic susceptibility testing can be another useful test for differentiating streptococci.

Using the previously discussed identification methods (hemolysis, morphology, biochemical, and antibiotic susceptibility), you will have an opportunity in this exercise to differentiate four streptococcus species: a Lancefield Group A species (*S. pyogenes*), a Lancefield Group B species (*S. agalactiae*), a Group D *Enterococcus*, and a nonenterococcus species.

Definitions

Beta-hemolysis.   A sharply defined, clear, colorless zone of hemolysis surrounding colonies of certain streptococci grown on blood agar plates.

Encapsulation.   A protective gelatinous material surrounding some bacteria that may also relate to their virulence.

Exotoxin.   A soluble, poisonous protein that passes into the growth medium during the growth of certain bacteria.

Glomerulonephritis.   Inflammation of the kidneys affecting the structure of the renal glomeruli (inflammation of the capillaries caused by toxins produced elsewhere in the body).

Inhibition zone.   The diameter in mm of no bacterial growth around an antibiotic disk. Bacteria were killed.

Precipitin reaction.   The reaction of an antibody with a soluble antigen to form an insoluble, visible antigen antibody complex.

Sequela.   An aftereffect of disease or injury that is often more serious than the initial disease.

Objectives

1. To determine semiquantitatively the number of *S. mutans*, *S. mitis*, and *S. salivarius* present in Snyder dental caries tubes.
2. To provide introductory information about the medical importance of the genus *Streptococcus* and how to identify members using a combination of morphological, biochemical (hemolysis, bile esculin, NaCl tubes, CAMP), and antibiotic susceptibility tests.
3. To determine the different Lancefield Groups—A, B, or D—with the bile esculin, NaCl, CAMP, and bacitracin antibiotic sensitivity tests.

References

Baron, E. J., Peterson, L. R, and Finegold, S. M. *Bailey and Scott's Diagnostic Microbiology*, 9th ed. Mosby, 1994.

Facklam, R. R., and Carey, R. B. 1985. "Streptococci and aerococci," 154–175. In E. H. Lennette, A. Balows, W. J. Hausler, Jr., and H. J. Shadomy (ed.), *Manual of clinical microbiology*, 4th ed. Washington, D.C.: American Society for Microbiology.

Lancefield, R. C. 1933. Serological differentiation of human and other groups of hemolytic streptococci. *J. Exp. Med.* 57:571–595. A classical paper worth examining.

Lindell, S. S., and Quin, P. 1975. Use of bile esculin for rapid identification of Enterobacteriaceae. *J. Clin. Microbiol.* 1:440–443.

Mahon, C. R., Lehman, D., Manuselis, G. *Textbook of Diagnostic Microbiology*, 3d ed. Elsevier Health Sciences, 2006.

National Committee for Clinical Laboratory Standards. Performance standards for antimicrobial susceptibility testing, 11th informational supplement. Wayne, PA: NCCLS, 2001.

Nester et al. *Microbiology: A human perspective*, 6th ed. The McGraw-Hill Companies, 2009. Chapter 17, Section 17.3; Chapter 22, Section 22.5; and Chapter 24, Section 24.3.

Materials per group

Fresh (24 hour, 37°C) unknown cultures of the following:

> *Enterococcus faecalis, Streptococcus bovis, Streptococcus pyogenes,* and *Streptococcus pneumoniae* on blood agar plates labeled 1, 2, 3, and 4.

Fresh (24 hour, 37°C) known cultures of the following:

> *Staphylococcus aureus* (must be β-hemolytic strain)
>
> *Streptococcus agalactiae*
>
> *Streptococcus pneumoniae*

Snyder dental caries tubes with 2–5 ml agar (held in a 60°C water bath, so still liquid)

Sterile empty petri dishes

Serological pipet (or micropipettor with tips) with a 0.5 ml (500 μl) capacity

Bile esculin slants or plates

TS broth with 6.5% NaCl

Blood agar plates

Bacitracin antibiotic disks

Optochin antibiotic disks

Ruler divided in mm

Forceps

Procedure

Safety Precautions: Because β-hemolytic streptococci are opportunistic pathogens (see exercise 23), all slides should be disposed of in the hazardous waste container. Any slants and plates should be promptly autoclaved. If any culture material is spilled, notify your instructor.

Dental Caries Test

1. Place 0.5 ml of saliva into a sterile petri dish. Immediately after removing the liquefied Snyder agar from the water bath, aseptically pipet 0.5 ml of saliva into the Snyder tube, and gently roll the agar tube between your hands to thoroughly mix the saliva into the agar. Allow this to solidify, and then label. Incubate at 37°C. Place used petri dishes and pipets or pipet tips into the proper disposal container.

2. Check for any color changes at 24, 48, and 72 hours. Record observations in the Laboratory Report.

Streptococcus Identification Testing

1. Prepare a gram stain of each unknown. Examine with the oil immersion objective lens and make drawings in the circled areas of the Laboratory Report. You are looking for an arrangement of pairs (diplococci) or chains indicative of streptococci.

2. For each unknown culture, use a sterile loop to obtain a colony, and inoculate a bile esculin slant or plate. Incubate at 37°C for 24–48 hours.

3. For each unknown culture, use a sterile loop to obtain a colony, and inoculate the TS broth containing 6.5% salt. Incubate at 37°C for 24–48 hours.

4. For each unknown culture, use a sterile loop to obtain a colony, and inoculate a blood agar plate with a lawn of growth. Be sure to record in the Laboratory Report the type of hemolysis seen for each unknown.

   a. With sterilized forceps, obtain a bacitracin antibiotic disk, and place on one-half of the blood agar plate. Incubate at 37°C for 24 hours.

   b. With sterilized forceps, obtain an optochin antibiotic disk, and place on the other half of the blood agar plate. Incubate at 37°C for 24 hours.

   c. Measure the inhibition zones of the antibiotic disks in mm. If the zone size is over 14 mm, the bacterium is considered susceptible.

CAMP Test

NOTE: Due to the pathogens involved, the instructor may elect to perform one CAMP test plate as a demonstration for the entire class.

1. Inoculate the middle of a blood agar plate with a straight streak of *Staphylococcus aureus*.

2. Perpendicular (right angle) to the *Staphylococcus* streak, inoculate with a straight streak of the *Streptococcus agalactiae*. The inoculated streak should meet with the *Staphylococcus* streak.

3. For a negative control, inoculate *S. pneumoniae* at the opposite end of the *Staphylococcus* streak. Incubate at 37°C for 24 hours. Record results in the Laboratory Report.

EXERCISE

28

Laboratory Report: Lancefield Grouping
of Pathogenic Streptococci

Results

1. For the Snyder dental caries test, describe the color of your tube at 24, 48, and 72 hours post-inoculation.

2. Drawings of unknown Gram-stained bacteria seen with the oil immersion objective lens:

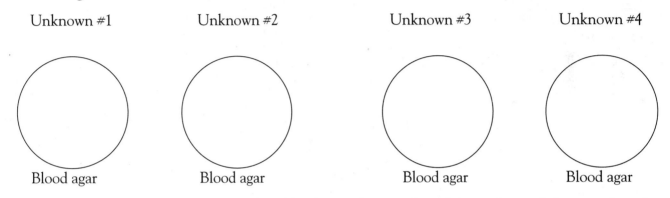

Unknown #1   Unknown #2   Unknown #3   Unknown #4

Blood agar   Blood agar   Blood agar   Blood agar

3. Describe the type of hemolysis found on blood agar, the growth on bile esculin, and the growth in the 6.5% NaCl tubes.

   Unknown #1:

   Unknown #2:

   Unknown #3:

   Unknown #4:

4. Record the unknown antibiotic inhibition zone results in table 28.2.

Table 28.2

Culture	Bacitracin	Optochin
1. Unknown #1		
2. Unknown #2		
3. Unknown #3		
4. Unknown #4		
5. Negative control		
6. Positive control		

5. From the three studies (morphology, hemolysis, and biochemical), which unknown did you identify as *S. pyogenes*? _____

   as *S. pneumoniae*? _____

6. Record the CAMP test results. Did you see an arrowhead indicative of *S. agalactiae*?

7. Were all of your findings consistent with the literature? If not, describe any inconsistencies observed, and, if possible, provide an explanation.

Questions

1. Why are Group D streptococci classified into the enterococcus and nonenterococcus groups?

2. What are some reasons for including positive and negative controls?

3. Why should antibiotic sensitivity testing be performed on β-hemolytic streptococci?

4. Why is *Streptococcus pneumoniae* not usually classified as a Lancefield Group when it possesses the cellular characteristics of Group B?

NOTES:

EXERCISE

Use of an Enzyme-Linked Immunosorbent Assay (ELISA) Test for *Coccidioides immitis* Identification

Getting Started

The ELISA test is widely used for identification of both plant and animal pathogens, including viruses. It is also used as an initial screening test for detecting antibodies to the AIDS virus. In the clinical setting, it is used to identify a variety of microbial pathogens because of test sensitivity and simplicity, often requiring only a swab sample from the infected host. It serves as an example of using a combined precipitin-enzyme reaction to achieve an end. Of great importance to the success of this technique is the plastic microwell plate, which can attract the reactants to its surface and hold on to them. In this exercise, the ELISA technique is used to identify a medically important dimorphic fungus, *Coccidioides immitis*, which assumes a yeastlike phase in the body of the host. In culture it produces a typical mold colony containing barrel-shaped arthrospores.

Various modifications of the ELISA antigen-antibody technique exist. The one used here is an indirect immunosorbent assay (figure 29.1*a*). With this method, the patient's antiserum is added to a microwell previously coated with a mixture of *Coccidioides immitis* antigens. If antibodies related to the coccidioidal antigens are present, they attach to the adsorbed antigens. After washing to remove unbound specimen components, an antibody conjugate that has coupled to the enzyme, horseradish peroxidase, is added. If binding occurs between the antigen and the antibody conjugate, a sandwich is formed, containing adsorbed antigens, patient antibodies, and the horseradish peroxidase enzyme. (Peroxidases are enzymes that catalyze the oxidation of organic substrates.) Next, the organic substrate used for this test, urea peroxide, is added. When oxidized by the peroxidase enzyme, free oxygen (O) is released. A color indicator, tetramethylbenzidine, is added, which when oxidized by the free oxygen, produces a yellow color. Lack of color means that the patient's antiserum does not contain *Coccidioides immitis* antibodies.

Another widely used ELISA method is the direct immunosorbent assay commonly known as the double antibody sandwich assay (figure 29.1*b*). It is widely used with plants to determine if they are infected with various pathogenic bacteria and viruses. Some of the symptoms include spotting and discoloration of the leaves and decreased yield. Some of these pathogens are able to infect a wide variety of plants. For example, tomato spotted wilt virus and impatiens necrotic spot virus are able to infect a variety of flowers and vegetables. Examples include hydrangeas, begonias, orchids, and potatoes.

References

Nester et al., *Microbiology: A human perspective*, 6th ed. The McGraw-Hill Companies, 2009. Chapter 22, Section 22.7.

Pappagianis, D., and Zimmer, B. L. 1990. "Serology of Coccidioidomycosis." *Clin. Microbiol. Reviews* 3:247–268.

Materials

A Meridian Bioscience, Inc.* Premier *Coccidioides* EIA Package Insert that contains the following:

Antigen coated microwells (96)—breakaway plastic microwells, each coated with a mixture of TP (tube precipitin) and CF (complement fixation) antigens.

Positive control (2.7 ml)—prediluted positive human serum with a preservative. Do not dilute further.

Note: The positive control serum and all materials that they contact should be handled at Biosafety Level 2 as recommended in the CDC/NIH manual "Biosafety in Microbiology

*We wish to thank Meridian Bioscience, Inc. for their help in providing for educational use Premier *Coccidioides* Enzyme Immunoassay test kits at a reasonable cost.

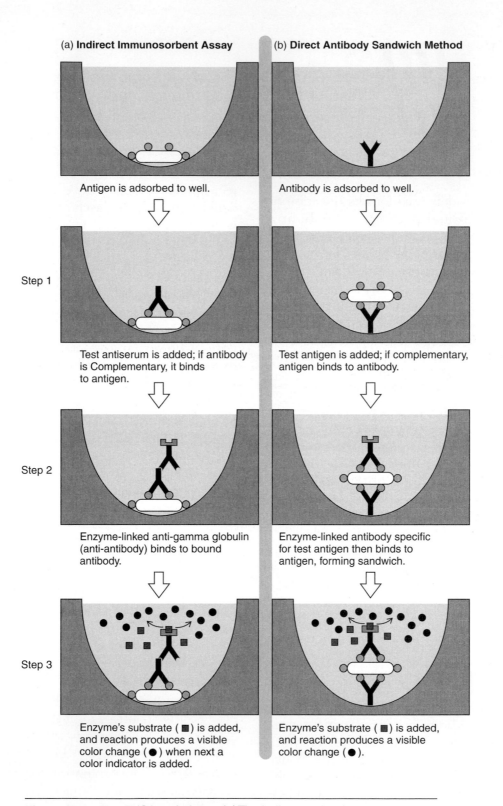

(a) Indirect Immunosorbent Assay

Antigen is adsorbed to well.

Step 1

Test antiserum is added; if antibody is Complementary, it binds to antigen.

Step 2

Enzyme-linked anti-gamma globulin (anti-antibody) binds to bound antibody.

Step 3

Enzyme's substrate (■) is added, and reaction produces a visible color change (●) when next a color indicator is added.

(b) Direct Antibody Sandwich Method

Antibody is adsorbed to well.

Test antigen is added; if complementary, antigen binds to antibody.

Enzyme-linked antibody specific for test antigen then binds to antigen, forming sandwich.

Enzyme's substrate (■) is added, and reaction produces a visible color change (●).

Figure 29.1 The ELISA technique. *(a)* The indirect antibody method, and *(b)* the double antibody sandwich method.

and Biomedical Laboratories," 1988. In view of using positive control serum, this exercise can only be conducted by personnel trained in handling pathogens. Laboratory instructors and students well grounded in proper use of aseptic technique should have no difficulty in meeting this requirement. The laboratory facilities must be at least Level 1, standard open bench, typical of most microbiology teaching laboratories.

Sample diluent (50 ml)—buffered protein solution with a preservative

20X wash buffer (50 ml)—concentrated wash buffer with a preservative

IgM enzyme conjugate (10 ml)—affinity purified goat anti-human IgM antibodies conjugated to horseradish peroxidase in buffered protein solution containing a preservative

IgG enzyme conjugate (10 ml)—affinity purified goat anti-human IgG antibodies conjugated to horseradish peroxidase in buffered protein solution containing a preservative

Horseradish peroxidase substrate (10 ml)—buffered solution containing urea peroxide and tetramethylbenzidine

Stop solution (10 ml)—2N sulfuric acid. **CAUTION:** Avoid contact with skin. Flush with water if contact occurs.

Microwell strip holder

Note: The expiration date is on the kit label. Store kit at 2°–8°C and return kit promptly to the refrigerator after use. Microwells not being used must be removed from the microwell holder and placed back inside the resealable foil pouch and sealed. It is important to protect the strips from moisture.

Other Materials

Pipets capable of delivering 10, 20, 100, and 200 μl

Test tubes (12 × 75 mm) for dilution of sample

Distilled or deionized water

Squirt bottle

Timer

If available, a plate reader capable of reading absorbance at 450 nm. A dual wavelength reader is preferred, using a second filter of 630 nm to correct for light scatter.

Note: The plate reader is not necessary for reading positive control results. The definite yellow color is easily observed with the naked eye.

Reagent Preparation

1 Bring the entire kit, including the microwell pouch, to 22°–25°C before and during use. Warming requires at least 1 hour.

2 Prepare 1X wash buffer by measuring 1 part of 20X buffer and diluting with 19 parts of water. The 1X buffer can be stored at room temperature for up to 1 month. Discard contaminated buffer. For performing washes, the buffer can be transferred to a wash, or "squirt," bottle.

Procedure

Note: Each microwell is coated with both TP (tube precipitin) and CF (complement fixation) antigens. It is recommended that both the IgM and the IgG antibody assays be performed simultaneously for the serum samples.

1. Snap off a sufficient number of microwells for the positive serum sample and the sample control and insert them into the microwell holder. Record sample positions, as shown in the example below.

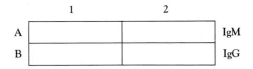

2. Add 100 μl of positive control antiserums (IgM and IgG) conjugated with horseradish peroxidase enzyme to each of two separate microwells (A1 and B1), and 100 μl of sample diluent (SD) to each of two separate microwells (A2 and B2).

Microwells A2 and B2 are negative controls. Any remaining wells will not be used in this exercise because they are for use with patient sera.

3. Mix the samples by gently rotating the microwell for 10–15 seconds on the countertop.

4. Immediately following step 3, incubate the microwell at 22°–25°C for 30 minutes.

5. Hold the microwell plate firmly from the bottom, and gently squeeze it.

   a. Dump plate contents into a biohazard receptacle. Next, strike the inverted plate firmly on a clean stack of paper towels or other absorbent material.

   b. With a wash bottle, fill all wells with 1X wash buffer by directing the stream of buffer to the sides of the wells to prevent foaming. Dump the plate contents in the biohazard receptacle, and strike the plate on the towels.

   c. Repeat step 5b two more times.

Note: The secret for success with this exercise lies in thorough rinsing of the microwells. After the final wash, strike the plate on fresh towels hard enough to remove as much excess wash buffer as possible, but do not allow the wells to completely dry at any time.

6. Add two drops of IgM enzyme conjugate to the microwells of row A.

7. Add two drops of IgG enzyme conjugate to the microwells of row B. Mix the samples by gently rotating the microwell plate for 10–15 seconds on the countertop.

8. Immediately following step 7, incubate at 22°–25°C for 30 minutes.

9. Repeat step 5, the washing cycle.

10. Add two drops of the urea peroxide and tetramethylbenzidene substrate to each microwell. Start the timer with the addition of substrate to the first well. Mix by gently rotating the microwells 10–15 seconds on the countertop.

11. Incubate at 22°–25°C for 5 minutes.

12. Carefully add two drops of stop solution (2N sulfuric acid) to each microwell in the same order as step 10. Mix as in step 10, and wait 2 minutes before reading. A yellow color on the bottom of the control serum wells (A1 and B1) represents a positive test. The bottom of the negative control wells (A2 and B2) should appear clear. Readings should be made within 15 minutes. Enter your results in table 29.1 of the Laboratory Report.

13. If a plate reader is available, you may be able to quantify your results. Carefully wipe the underside of the microwells with a lint-free tissue, and measure the absorbance at 450 nm as follows:

    a. Blank on air.

    b. Read the negative control microwells; values should be <0.100.

    c. Then reblank the reader using the negative control well values.

    d. Enter your results in table 29.2 of the Laboratory Report.

14. Disinfect and retain microwell holder. Discard used assay materials in biohazard container.

Note: The positive and negative controls must be assayed with each batch of patient specimens to provide quality assurance. The negative control (sample diluent) should yield an A 450 value <0.100 with both IgM and IgG conjugates when blanked on air. The positive control microwells should have a definite yellow color and yield an A 450 value >0.500 and <2.500 with both IgM and IgG conjugates.

For patient specimens, the results have been interpreted as follows:

Negative = Absorbance Value <0.150
Indeterminate = Absorbance Value >0.50 but <0.199
Positive = Absorbance Value >0.200

Interpretation of Results

A negative result with both IgM and IgG indicates that serum antibody to *Coccidioides immitis* antigens is either absent, below the level of detection of the assay, or the specimen was obtained too early in the response. A positive result with either IgM or IgG implies the presence of antibody to *C. immitis*. A positive response with either conjugate should be reported. An early acute-phase patient may only present an IgM response, while the chronic or convalescent patient may only present an IgG response.

Specimens that yield an indeterminate result should be retested. If the retest result is still indeterminate, a second specimen should be obtained. Extremely strong positive reactions may yield a purple precipitate. Absorbances obtained with such reactions may be lower than expected but are still positive.

Limitations of the Procedure

A negative result with both IgM and IgG antibodies does not preclude diagnosis of coccidioidomycosis, particularly if only a single specimen has been tested and the patient shows symptoms consistent with a positive diagnosis.

A positive ELISA result should be confirmed by an Ouchterlony immunodiffusion assay (see exercise 30). However, because of the relative insensitivity of the ID (immunodiffusion) procedures, an ID-negative test does not prevent the possibility of coccidioidomycosis. Diagnosis is based on both laboratory and clinical findings, as well as the presence of antibody.

Results should be compared with patient symptoms to determine if there is a logical correlation.

NOTES:

EXERCISE

29

Laboratory Report: Use of an Enzyme-Linked Immunosorbent Assay (ELISA) Test for *Coccidioides immitis* Identification

Results and Questions

1. Discuss the test results and their significance.

Table 29.1 Visual Observation of Processed Samples

Sample	Yellow Color (I)
IgM	
IgM control	
IgG	
IgG control	

Table 29.2 Sample Absorbance at 450 nm with a Plate Reader

Sample	Absorbance
Negative Control	
IgM	
IgG	

2. Discuss laboratory safety considerations related to handling of

   a. antigen-coated microwells.

   b. positive serum control.

   c. immunoglobulin enzyme conjugates.

   d. urea peroxide.

   e. tetramethylbenzidine.

3. What is the importance of rinsing when conducting the ELISA test?

4. Why is the ELISA test for coccidioidomycosis more definitive than the Ouchterlony immunodiffusion test (exercise 30)? Is the latter test still of value as a diagnostic tool? Explain your answer.

5. Discuss the pros and cons of using the enzyme-linked immunosorbent assay (ELISA) and the double antibody sandwich ELISA assay. You may need to consult your text for the answer to this question.

EXERCISE 30

An Ouchterlony Double Immunodiffusion Test for *Coccidioides immitis* Identification

Getting Started

Precipitin reaction tests, such as the Ouchterlony test, are widely used for **serodiagnosis** of fungal diseases. More recently, the ELISA test (see exercise 29) has also been used for this purpose. It is not unusual to use two or more serological tests for initial diagnosis of a fungal disease. The latex slide agglutination test (see exercise 28) has also been used as an initial diagnostic test for fungal disease.

With the Ouchterlony procedure, soluble antigen as well as **serum** solutions containing antibodies are placed in separate wells of an agar base. The reactants diffuse from the wells and form a thin, white precipitin line(s) where they meet in optimum proportions (figure 30.1a). Since both immunoreactants move in this system, it is known as a **double immunodiffusion test.** In a related system, the Oudin technique, diffusion occurs only in a single dimension (figure 30.1b).

According to the **lattice theory** depicted in figure 30.2, the precipitin lines form best where an excess of antibody relative to antigen is present. When the Ab/Ag ratio is less than 1, soluble complexes, rather than precipitates, occur (figure 30.2d–f). In some instances, more than one precipitin line forms in the agar because the antigen preparation contains more than one type of antigen; thus, each band formed represents one antigen-antibody reaction.

The curvature of the precipitin line provides information about the molecular weight of the antigen, providing the antigen and antibody are present in nearly equal amounts. The precipitin line appears straight if the antigen and antibody have about the same molecular weight. If the antigen has a higher molecular weight, the line is concave toward the antibody well; if the antigen has a lower molecular weight, the line is concave toward the antigen well. These relationships are derived from the principle that the rates of diffusion of any

Figure 30.1 Some commonly used systems for gel diffusion precipitin reactions. *(a)* Double diffusion in two dimensions (Ouchterlony technique), in which diffusible antigen and antibody solution are placed in separate wells cut in an agar plate. (Direction of diffusion is shown by arrows.) *(b)* Single diffusion in one dimension (Oudin technique) using a soluble diffusion antigen layered over an antibody contained in an agar gel column. Black lines are opaqued precipitin bands. Reproduced with the permission of Meridian Bioscience, Inc., Cincinnati, Ohio 45244.

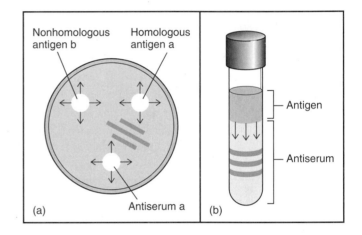

mixture increase with concentration and decrease with molecular weight.

Finally, antigens or antibodies can be tested for identity by placing a test well of the substance in question adjacent to the wells of a known system (figure 30.3). If the antigen-antibody complexes are identical, the precipitin lines form an unbroken line of identity with the known system (figure 30.3a). Partial identity and nonidentity reactions are also possible. A partial identity (figure 30.3b) reaction occurs when certain components of the antigens (or antibodies) are identical and others are not. The spur represents the components that are unrelated. A nonidentity reaction (figure 30.3c) occurs when the antigen-antibody complexes are different. The resulting "X," or cross reaction, indicates that two unrelated complexes are present.

Figure 30.2 Use of the lattice theory to postulate the proper ratios of Ab/Ag necessary to form a precipitin line.

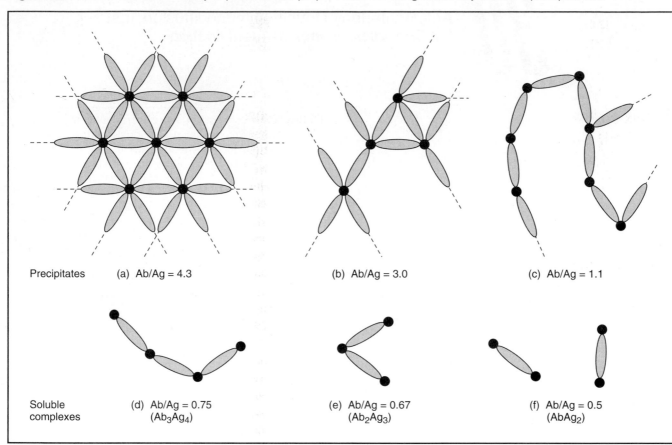

Precipitates   (a) Ab/Ag = 4.3   (b) Ab/Ag = 3.0   (c) Ab/Ag = 1.1

Soluble   (d) Ab/Ag = 0.75   (e) Ab/Ag = 0.67   (f) Ab/Ag = 0.5
complexes   (Ab_3Ag_4)   (Ab_2Ag_3)   ($AbAg_2$)

The test organism chosen to illustrate principles of Ouchterlony double immunodiffusion is the sometimes pathogenic mold *Coccidioides immitis*. In nature it survives best in desert soils where the temperatures average 100°F in the summer and 33°–38°F in the fall and winter. It actually grows better on rich soil than on poor. However, survival is greatly reduced on soil containing normal bacteria and fungal flora. Thus, the most highly endemic regions for contracting coccidioidomycosis in North America are the southwestern United States and northern Mexico. In 1892, when first examined in the lesions of infected patients, the morphology of the endosporulating spherule suggested a relation to protozoa in the order Coccidia. Later laboratory studies establishing the same endospore formation within spherules implied a relationship to the fungal class Zygomycetes. We now know that *Coccidioides immitis*

is a dimorphic yeastlike fungus belonging to the class Ascomycetes. Nevertheless, the genus name *Coccidioides* remained. The species name, *immitis*, means not mild. Some common names for this disease are San Joaquin Valley fever, Valley bumps (small tender reddened nodules under the skin), and California disease.

Rather than a live fungus for an antigen source, a purified culture filtrate of *Coccidioides immitis* containing the F antigen at a concentration of 100 units/ml will be used in this study. The antiserum source was produced in hyperimmune (not infected) goats injected with the purified F antigen.

An interesting alternative method for studying the Ouchterlony procedure is one that uses various unknown meat samples as an antigen source and anti-horse, pig, and cow albumin samples for antisera (antibodies).

Figure 30.3 Double diffusion precipitin reactions observed in agar gel plates.

1. Antigen solutions with one or more antigenic components. Antigen solution A contains two distinct antigen components, as indicated by precipitin lines a_1 and a_2.
2. Lines of identity (fusion). *(a)* Antigen A solution contains a component antigenically identical with antigen solution B. For example, A shows a line of identity by fusion and thickening of precipitin lines a_2 with b_2.
3. Lines of partial identity (spur). *(b)* Antigen C contains a component (c_2) that shares at least one antigenic determinant with antigen A, but A contains at least one antigenic determinant that is nonidentical to C. The evidence for the partial identity of antigens A and C is the merging and thickening of the C precipitin line with the A precipitin line, but the A precipitin line has a spur that overlaps the C precipitin line.
4. Lines of nonidentity (cross). *(c)* In this instance, the precipitin lines do not fuse but cross because the antigens A and D are not related chemically to one another.

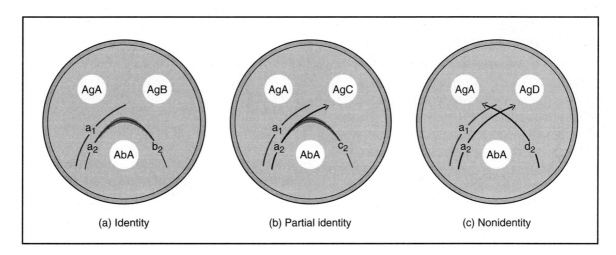

(a) Identity   (b) Partial identity   (c) Nonidentity

Definitions

Double immunodiffusion test. An immunological test in which both the antigen and the antibody move in the reaction system.

Lattice theory. A theory based on the framework formed when soluble antigens and antibodies cross in a diagonal manner. When the correct ratio of Ab to Ag is present, an aggregate forms. When the aggregate is sufficiently large, it precipitates out of solution in the form of a visible complex.

Precipitin reaction. The reaction of an antibody with a soluble antigen that results in an insoluble precipitate.

Precipitin reaction test. A test in which an antibody combines with an identifiable soluble antigen, causing aggregation and eventual precipitation out of solution.

Serodiagnosis. A diagnostic test for antigen-antibody reactions that employs the serum portion of the blood.

Serum. The fluid portion of the blood that remains after the blood clots. Contains antibodies, hormones, and dissolved nutrients but not cells.

Objectives

1. To familiarize you with some of the principles of Ouchterlony double immunodiffusion, and to provide an opportunity for you to conduct and observe some of these principles at work in the laboratory.
2. To suggest some improvements in methodology once you have conducted this exercise.

Reference

Nester et al. *Microbiology: A human perspective,* 6th ed. The McGraw-Hill Companies, 2009. Chapter 22, Section 22.7.

Materials

A Meridian Bioscience, Inc. Test Kit (603096)* that contains:

0.5 ml *Coccidioides* ID antigen

Diluted anti-*Coccidioides* ID control serum

Immunodiffusion agar plates

Capillary pipets with bulb

Moist chamber: a dish with a tight-fitting cover containing moist paper toweling is satisfactory, provided the ID plates remain level and hydrated during the incubation period.

An inexpensive reading light, such as a desk lamp with a black cover that contains a 60-watt lightbulb for illumination. By holding the Ouchterlony plate vertically near the lower outside edge of the black cover, you should be able to see well-defined precipitin lines.

Reagent quality or distilled water

*We wish to thank Meridian Bioscience, Inc. for their willingness to provide short-dated, usable immunodiffusion kits at a reasonable cost.

Procedure

Note: Refer to figure 30.4 of the Laboratory Report for a description and numerical designation of the test well pattern of the ID agar plate.

1. Using a capillary pipet with an attached bulb, fill the pipet approximately three-quarters full with *Coccidioides* antigen.
2. Next, fill the center well (#7 on figure 30.4) with *Coccidioides* antigen.

    Note: Take care to avoid overflow of antigen solution from the well.
3. Using a fresh capillary pipet, repeat steps 1 and 2 with positive control antiserum. Fill well #1 with positive control serum. If you wish to observe a line of identity, also fill well 2 or 6 with positive control serum (see color plate 17).

4. If negative control sera or positive sera from patients known to have a *Coccidioides* infection is available, it can be added to other external wells of the ID agar plate. For example, if positive serum is available from a patient, it should be placed in a well adjacent to one containing positive control antiserum. Such an arrangement will enable you to determine lines of identity, partial identity, or nonidentity (see figure 30.3 and color plate 17).
5. Identify the contents of each well in table 30.1 of the Laboratory Report. Also make a note of the wells, if any, in which the antiserum was diluted, and the amount of the dilution.
6. Place your name or initials and date on the ID plate cover. Incubate it in the moist chamber at room temperature for 24–48 hours.
7. After 24–48 hours' incubation, read and record the ID bands in table 30.1 of the Laboratory Report. A light source is preferred for observing the nature of the bands (see the Materials section). Particular attention should be paid to the orientation of the bands in relation to control serum bands. A smooth junction of the bands is indicative of an identity reaction (see color plate 17). If antiserum from a patient was included in this test, you should also look to see if the control band is bent toward a position in front of the patient well. This indicates patient antibody at a low titer.
8. Interpretation of the test.
    a. A band of identity with a known positive control indicates the presence of patient antibody against the antigen in question. In general, an identity reaction against a given antigen is indicative of active or recent-past infection.
    b. Partial identity reactions are regarded as positive for antibody against the antigen only if no other identity reaction is present on the plate. Partial identity reactions are also indicators of possible disease.

c. Nonidentity reactions are regarded as a negative test; they may also be apparent when the disease state is caused by a mycotic agent other than the one tested.

d. The greater the number of precipitin lines observed, the greater the likelihood of severe disease. In arriving at a diagnosis, the test results should be interpreted with all other lab and clinical data, including treatment history.

9. Limitations of the test. The high rate of negative serologic tests observed among culturally demonstrable cases limits the predictive value of a negative test. Positive test results must be confirmed culturally.

NOTES:

Name _____ Date _____ Section _____

EXERCISE

30

Laboratory Report:
An Ouchterlony Double Immunodiffusion Test
for *Coccidioides immitis* Identification

Results

Date ID plate incubated ___ Date observed ___

1. Make drawings of all observed precipitin lines on figure 30.4.

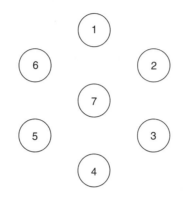

Figure 30.4 Test well pattern of
the immunodiffusion agar plate.

2. Fill in the necessary blanks in table 30.1.

Table 30.1   Ouchterlony Fungal Immunodiffusion Analysis Form

Well No.	Well Description	Reading Observed
1	Positive control serum	
2		
3		
4		
5		
6		
7	*Coccidioides* antigen:	

Note: Indicate under "Well Description" in which wells, if any, the serum was diluted.

Note: Test is invalid if positive control is negative after a 24-hour incubation.

3. Discuss your results and their significance.

4. What future studies do your results suggest?

Questions

1. Interpret the following *Coccidioides* immunodiffusion bands with respect to the possibility of having an active *Coccidioides* infection:

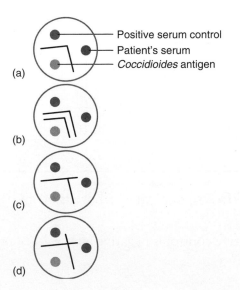

(a) Positive serum control — Patient's serum — *Coccidioides* antigen

(b)

(c)

(d)

2. Interpret the relative antigen (Ag) and antibody (Ab) molecular weights from the following immunodiffusion bands:

   a. ID test using equal concentrations of Ag and Ab.

   o – – Ab

   ————

   o – – Ag

   b. ID test using a greater concentration of Ag than Ab.

   o – – Ab

   ————

   o – – Ag

   c. ID test using equal concentrations of Ag and Ab.

   o – – Ag

   o – – Ab

3. Circle the Ag (•) Ab (⬤) diagram in which the best precipitin line is possible:

4. What advantages does Ouchterlony double immunodiffusion precipitin analysis have over Oudin single immunodiffusion analysis?

NOTES:

*I*NTRODUCTION

to the Prevention and Control
of Communicable Diseases

Communicable, or infectious, diseases are transmitted from one person to another. Transmission is either by direct contact with a previously infected person—for example, by sneezing—or by indirect contact with a previously infected person who has contaminated the surrounding environment.

A classic example of indirect contact transmission is an epidemic of cholera that occurred in 1854 in London. During a 10-day period, more than 500 people became ill with cholera and subsequently died. As the epidemic continued, John Snow and John York began a study of the area and were able to prove by **epidemiological methods** only (the bacteriological nature of illness was not known at that time) that the outbreak stemmed from a community well on Broad Street known as the Broad Street Pump* (figure I.10.1). These researchers then discovered that sewage from the cesspool of a nearby home was the pollution source and that an undiagnosed intestinal disorder had occurred in the home shortly before the cholera outbreak. They also learned that neighboring people who abstained from drinking the pump water remained well, whereas many of those who drank the water succumbed to cholera.

Today the incidence of cholera, typhoid fever, and other infectious diseases rarely reaches **epidemic** proportions in those countries that have developed standards and regulations for control of environmental reservoirs of infection, particularly the major reservoirs: water, food, and sewage. (The subject of public health sanitation is presented in exercise 32, which discusses water microbiology.)

All public health sanitation studies come under the surveillance of public health agencies responsible for prevention and control of communicable diseases. Among these are the federal cabinet-level Department of Health and Human Services, which conducts preventive medicine research, provides

hospital facilities for military personnel, and gives financial assistance to state and local health departments, as well as assistance at times to developing countries. Additionally, the Centers for Disease Control (CDC) in Atlanta, plays an important role in the prevention and control of communicable diseases. All state and local government agencies also perform important public health services.

Perhaps the smost important international health agency is the World Health Organization (WHO), headquartered in Geneva, Switzerland. WHO distributes technical information, standardizes drugs, and develops international regulations important for the control and eradication of epidemic diseases. For example, smallpox, which was once a widespread disease, is virtually nonexistent today. Its current emphasis is on preventing HIV disease in mothers and children.

Finally, there are voluntary health organizations that help in some of the causes previously mentioned. A notable example is the Bill and Melinda Gates Foundation which focuses on fighting diseases in developing countries.

Definitions

Epidemic.   The occurrence in a community or region of a group of illnesses of similar nature, clearly in excess of normal expectancy.

Epidemiological methods.   Methods concerned with the extent and types of illnesses and injuries in groups of people and with the factors that influence their distribution. This implies that disease is not randomly distributed throughout a population, but rather that subgroups differ in the frequency of different diseases.

---

*Snow, John. "The Broad Street Pump," in Roueche, Berton (ed.), *Curiosities of medicine*. New York: Berkely, Medallion, ed., 1964.

Figure I.10.1 *(a)* The John Snow pub in London where epidemiologists go to celebrate the heroics of John Snow's early epidemiological efforts to stem a cholera epidemic. *(b)* A replica of the pump, with pump handle attached, a monument dedicated to Dr. Snow in July 1992. At the time of the epidemic, he was so convinced that the disease was being carried by water from the pump that he had the pump handle removed. Koch isolated and identified the cholera vibrio about 30 years later. Courtesy of Kathryn Foxhall.

(a)

(b)

31

What is the Role of *Propionibacterium acnes* in Acne?

Getting Started

In this exercise, we will examine the question of how the presence of the bacterium *Propionibacterium acnes* on the skin correlates with the symptoms of **acne**. Acne is a skin disease characterized by pimples, blackheads, and sometimes deeper eruptions. It is very common: 80% of people ages 11–30 experience it at least once. It rarely affects people in their thirties and forties.

What is known about acne, in brief, is that sebum, or oil, is secreted from the sebacious glands through pores in the skin (figure 31.1). If the pore becomes clogged, pimples and blackheads form and the associated hair follicle becomes infected. It is most common on the face and back.

When this disease was first studied, a Gram-positive, anaerobic **diphtheroid** was almost always cultured from the patient's skin. Therefore, this organism was considered the cause of acne and named ***Propionibacterium acnes***. Later the skin of persons who had no symptoms of acne was also examined, and surprisingly, *P. acnes* could also be cultured from their skin, sometimes in very large numbers. This suggested that *P. acnes* did not have a role in producing acne. However, there is some evidence that *P. acnes* may be actively involved in causing some of the symptons. The bacteria may break down lipid skin secretions into irritating components or may activate receptor sites of certain cells in the immune system.

The question, then, is what role does *P. acnes* play, if any, in the skin lesions of acne? When attempting to answer this question, the first step is to repeat the initial observations of the presence of *P. acne* on normal skin. This is extremely important because it is fairly easy to have a set of result published, but it is essential that the results can be repeated by others.

In this exercise, we will try to support the observation that *P. acnes* is a component of almost everyone's skin biota, whether acne is present or not. Students will sample their own foreheads for the presence of *P. acnes*. This procedure will be somewhat quantitative. The student can anonymously give the results of his or her forehead sample and indicate whether he or she has acne. The instructor can compile the class results and discuss how the presence of *P. acnes* and the condition correlates in this very small sample.

Definitions

Acne. A skin condition characterized by whiteheads, blackheads, and infected hair follicles.

Diphtheroids. Gram-positive, irregularly club-shaped rods.

Objectives

1. To gain an understanding of the scientific approach to problem solving.
2. To gain an appreciation of the difficulties of finding the cause of acne.

References

Evans, Charles. Persistent individual differences in the bacterial flora of the skin of the forehead: numbers of *Propionibacteria*. *The Journal of Investigative Dermatology*, 64:42–46, 1975.
MayoClinic.Com
Nester et al. *Microbiology: A Human Perspective*, 6th ed. The McGraw-Hill Companies, 2009.

Materials

Per Student
  TSY agar plates, 2
  5.0 ml sterile saline, 1
  9.9 ml sterile saline blank, 1

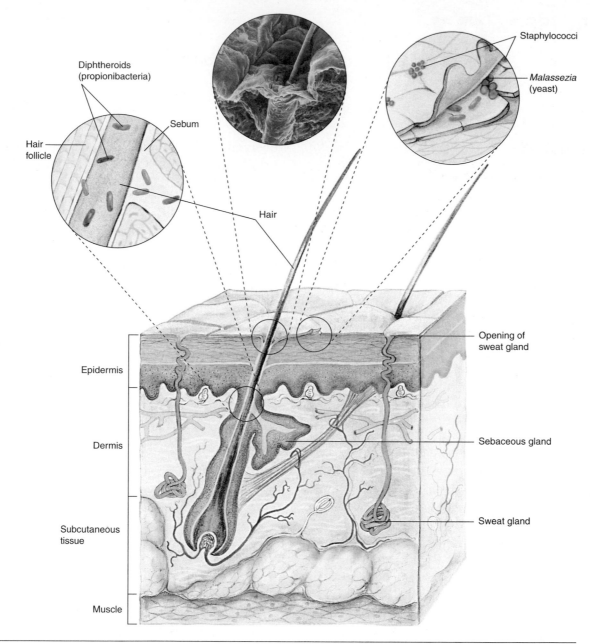

Diptheroids
(propionibacteria)

Sebum

Hair
follicle

Hair

Staphylococci

Malassezia
(yeast)

Epidermis

Dermis

Subcutaneous
tissue

Muscle

Opening of
sweat gland

Sebaceous gland

Sweat gland

Figure 31.1   Diagram of a hair follicle.

sterile swab, 1

1 ml sterile pipets, 3

sterile bent glass rods (sometimes called
dally rods or hockey sticks), 2

cotton ball

70% ethanol

hydrogen peroxide for catalase test

Procedure

First Session

1. Swab your forehead with cotton soaked in
   70% ethanol.
2. Permit the skin to dry without touching the
   forehead for 30 minutes.
3. Sample partner's forehead. Moisten the swab
   in the sterile saline. Carefully swab an area
   about the size of a quarter for 15 seconds, and

then rinse the swab in the tube of saline. Discard the swab as directed. Mix the sample thoroughly.

4. Dilute the sample 1/100 by adding 0.1 ml of the bacterial suspension to the 9.9 ml saline blank. Mix thoroughly. Remove 0.1 ml of the diluted sample, and place it on the surface of the TSY agar plate. Immediately spread it over the surface of the agar with a sterile bent glass rod. Place the rod in a container to be washed and sterilized as directed by the instructor.

5. Repeat the procedure with the undiluted sample.

6. Invert the plates, and incubate in an anaerobe jar at 37°C for 4–5 days.

Second Session

1. Examine the plates for small (1 mm) round raised colonies, and prepare a Gram stain to verify that the colonies are *Propionibacterium acnes*. They should appear as Gram-positive club-shaped rods (see figure 22.1). Use part of the same colony if possible to perform a catalase test (figure 23.3 a). Although *Propionibacteria* are obligate anaerobes, they are catalase positive.

2. A variety of organisms are found on the skin. In addition to *Propionibacteria* you may observe the following bacteria. *Staphylococcus epidermidis* colonies are large, flat, and white or gray. They are consistently found on skin.

Propionibacterium granulosum colonies are larger than *P. acnes*, buff colored instead of white, and usually present in small numbers. These are not important for this study. *Staphylococcus aureus* colonies are large and yellow. **These can be pathogens and should be handled with extra care**.

3. Estimate the number of *Propionibacteria acnes* on the plate that has the most countable colonies. If the numbers are very large, divide the plate in fourths and count one section. Record as TNC (too numerous to count) if you cannot distinguish separate colonies. Calculate the estimated number of *Propionibacteria acnes* from the swabbed area.

4. On a separate piece of paper, provide the estimated number of *P. acnes* from the area sampled and indicate if you now have acne or are being treated for acne (yes or no). Hand the paper into the instructor so that the data can be posted. Please do not sign your name to the paper.

5. Record the results.

6. Discuss if the class results support the original observation: *P. acnes* is a component of almost everyone's skin biota, whether acne is present or not.

7. It is important to note that these results can only be described as preliminary because the sample size is very small. Also, for a study to have validity, the techniques must be carefully standardized.

NOTES:

EXERCISE 31

Laboratory Report:
What Is the Role of *Propionibacterium acnes* in Acne?

Results

Individual

How many P. *acnes* did you isolate from the swabbed area of your forehead?

Class

Number of students in the class. _____
Number of students isolating P. *acnes* from his or her forehead. _____
Number of students with self-described acne. _____
Number of students with large numbers of P. *acnes* and acne. _____
Number of students with large numbers of P. *acnes* and no acne. _____
Conclusion:

Questions

1. What structure of the skin is involved in acne?

2. What two characteristics distinguish propionibacteria from other organisms found on the skin?

3. Why should initial experiments be repeated before further study?

4. At one time, greasy foods and chocolate were thought to cause acne, but these have been ruled out. How would you design a study to verify this? Consider that persons in the age group most prone to acne probably eat a substantial amount of these foods.

NOTES:

EXERCISE

32

Bacteriological Examination of Water: Multiple-Tube Fermentation and Membrane Filter Techniques

Getting Started

> Water, water, everywhere Nor any drop to drink.
> *The Ancient Mariner,* Coleridge, 1796

This rhyme refers to seawater, undrinkable because of its high salt content. Today the same can be said of freshwater supplies, polluted primarily by humans and their activities. A typhoid epidemic, dead fish on the beach, and red tides are all visible evidence of pollution. There are two primary causes of water pollution: dumping of untreated (raw) sewage and inorganic and organic industrial wastes, and fecal pollution by humans and animals of both freshwater and groundwater. In the United States, sewage and chemical wastes have been reduced largely as a result of federal and local legislation requiring a minimum of secondary treatment for sewage as well as severe penalties for careless dumping of chemical wastes.

Fecal pollution by humans and animals is more difficult to control, particularly as the supply of water throughout the world becomes more critical. In some parts of the world, particularly in developing countries, it is estimated that over 12,000 children die every day from diseases caused by waterborne fecal pollutants. Examples of such diseases are cholera, typhoid fever, bacterial and amoebic dysentery, and viral diseases such as polio and infectious hepatitis. Most of the inhabitants of these countries are in intimate contact with polluted water because they not only drink it, but they also bathe, swim, and wash their clothes in it.

The increased organic matter in such water also serves as a substrate for **anaerobic bacteria,** thereby increasing their numbers in relation to the **aerobic bacteria** originally present (figure 32.1). Nuisance bacteria, such as *Sphaerotilus natans*, a large, rod-shaped organism that grows in chains and forms an external sheath (figure 32.2), are able to adhere to

Figure 32.1 Diversity of microbes found in pure to very polluted water. Note the change from aerobic to anaerobic microbes as the water becomes more polluted. Courtesy of Settlemire and Hughes. *Microbiology for Health Students,* Reston Publishing Co., Reston, Virginia.

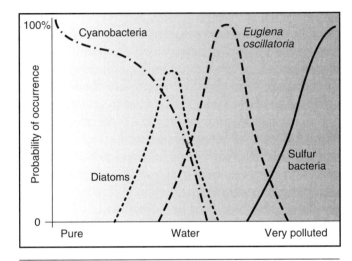

Figure 32.2 *Sphaerotilus* species, a sheathed bacterium that often produces masses of brownish scum beneath the surface of polluted streams. Phase-contrast photomicrograph courtesy of J. T. Staley and J. P. Dalmasso.

the walls of water pipes. This can eventually cause fouling of the pipes, thereby reducing their carrying capacity.

Fortunately, microbes are also beneficial in water purification. In smaller sewage treatment plants, raw sewage is passed through a slow sand filter, wherein

Figure 32.3 Human fecal specimen illustrations showing forms of *Giardia lamblia,* a waterborne protozoan pathogen that inhabits the intestinal tract of certain warm-blooded animals. *(a)* Pear-shaped trophozoite. Electron microscopy, iodine-stained, magnification × 2,900, 1 micron = 2.9 mm. Note flagellate appendages for locomotion. *(b)* Smooth ovoid cysts. The cyst form is resistant to adverse environments and is the form released with fecal material. Scanning electron microscopy, magnification × 2,900, 1 micron = 2.9 mm. Cysts are embedded in a mat of debris, bacteria, and fecal material. *(a)* © J. Paulin/Visuals Unlimited *(b)* From D. W. Luchtel, W. P. Lawrence, and F. B. De Walle, "Electron Microscopy of Giardia lamblia Cysts," *Applied and environmental microbiology,* 40:821–832, 1980. © American Society for Microbiology.

Drawing of trophozoite

microorganisms present in the sand are able to degrade (metabolize) organic waste compounds before the effluent is discharged. Sewage effluent is finally chlorinated to further reduce fecal microbial contaminants. The development of sewage treatment plants and the control of raw sewage discharge reduced the annual typhoid fever death rate in the United States from about 70 deaths per 100,000 population to nearly zero. However, the potential danger of pollution is always present. In 1973, an epidemic outbreak of typhoid fever occurred in Florida, and in 1975, residents of the city of Camas, Washington, were inundated by intestinal disorders that were traced to fecal pollution of the water supply by beavers infected with the protozoan *Giardia lamblia* (figure 32.3).

Two microbiological methods are commonly used to determine whether a given sample of water is polluted:

1. Total number of microorganisms present in the water. The plate count method provides an indication of the amount of organic matter present. In view of the great diversity in microbial physiology, no single growth medium and no single set of cultural conditions can be found that satisfy universal microbial growth. Hence, a choice has to be made. Experience has taught that standard plate counts on nutrient agar at two incubation temperatures, 20°C and 35°C, provide a useful indication of the organic pollution load in water.

2. **Indicator organisms.** Indicator organisms are normally nonpathogenic, always occur in large quantities in feces, and are relatively easy to detect compared to waterborne pathogens. The diagnosis of pathogens is

(a)

(b)

usually more complicated and time-consuming, and thus less suited for routine investigations. Assuming that in cold surface waters the pathogens are dying off faster than the indicator organisms employed, the absence or presence in very low numbers of the latter guarantees, in most cases, the absence of pathogens. Recently, however, better techniques and procedures have been developed for the detection of pathogenic bacteria and viruses. In certain cases, these techniques have indicated the presence of pathogens in the absence of indicator organisms hitherto relied upon, showing the pathogens' increased resistance to the aqueous environment. Similar findings have been found with respect to chlorination resistance. Such results suggest that some of the principles and methods in conventional water examination are of questionable value.

Despite their limitations, conventional methods for detection of fecal contamination are still widely employed. Some will be described here. In general, these procedures are the same ones described by the American Public Health Association in *Standard Methods for the Examination of Water and Wastewater*.

The indicator organisms most widely used are **coliform bacteria.** These include all aerobic and facultative anaerobic, Gram-negative, non-spore-forming, rod-shaped bacteria that ferment lactose with gas formation within 48 hours at 37°C, and comprise *Escherichia coli* (10^6 to 10^9 cells/g of feces) together with a number of closely related organisms (see exercise 24). Noncoliforms that are sometimes employed, primarily for confirmation, include *Streptococcus faecalis*, some related species, and in Great Britain, *Clostridium perfringens*, which is also called *Clostridium welchii*.

The presence of *E. coli* in water from such sources as reservoirs suggests that chlorination is inadequate. Current standards for drinking water state that it should be free of coliforms and contain no more than ten other microorganisms per ml.

Two of the most important methods applied to detect coliform organisms are the multiple-tube fermentation technique and the membrane filter technique.

Multiple-Tube Fermentation Technique

This technique employs three consecutive tests: first, a presumptive test; if the first test is positive, then a confirmed test; and, finally, a completed test (figure 32.4).

Presumptive Test   This test, a specific enrichment procedure for **coliform bacteria,** is conducted in fermentation tubes filled with a **selective growth medium** (MacConkey lactose broth), which contain inverted Durham tubes for detection of fermentation gas (see figure 32.4).

The main selective factors found in the medium are lactose; sometimes a **surfactant,** such as Na-lauryl sulfate or Na-taurocholate (bile salt); and often a pH indicator dye for facilitating detection of acid production, such as bromcresol purple or brilliant green. The selective action of lactose occurs because many bacteria cannot ferment this sugar, whereas coliform bacteria and several other bacterial types have this ability. The surfactant and dye do not inhibit coliform bacteria, whereas many other bacteria, such as the spore formers, are inhibited. The original surfactant used by MacConkey was bile salt because *E. coli* is adapted to growing in the colon, whereas nonintestinal bacteria generally are not.

The formation of 10% gas or more in the Durham tube within 24–48 hours, together with turbidity in the growth medium, constitutes a positive presumptive test for coliform bacteria and hence for the possibility of fecal pollution. The test is presumptive only; under these conditions several other types of bacteria can produce similar results.

The presumptive test also enables quantitation of the bacteria present in the water sample. The test, the most probable number test (MPN), is also useful for counting bacteria that reluctantly form colonies on agar plates or membrane filters but grow readily in liquid media. In principle, the water sample is diluted so that some of the broth tubes contain a single bacterial cell. After incubation, some broth tubes show growth with gas, whereas others do not. The total viable count is then determined by counting the portion of positive tubes and referring this data to a statistical MPN table used for calculating the total viable bacterial count (see Procedure, table 32.1).

Figure 32.4 Standard methods procedure for the examination of water and wastewater and for use in determining most probable number (MPN).

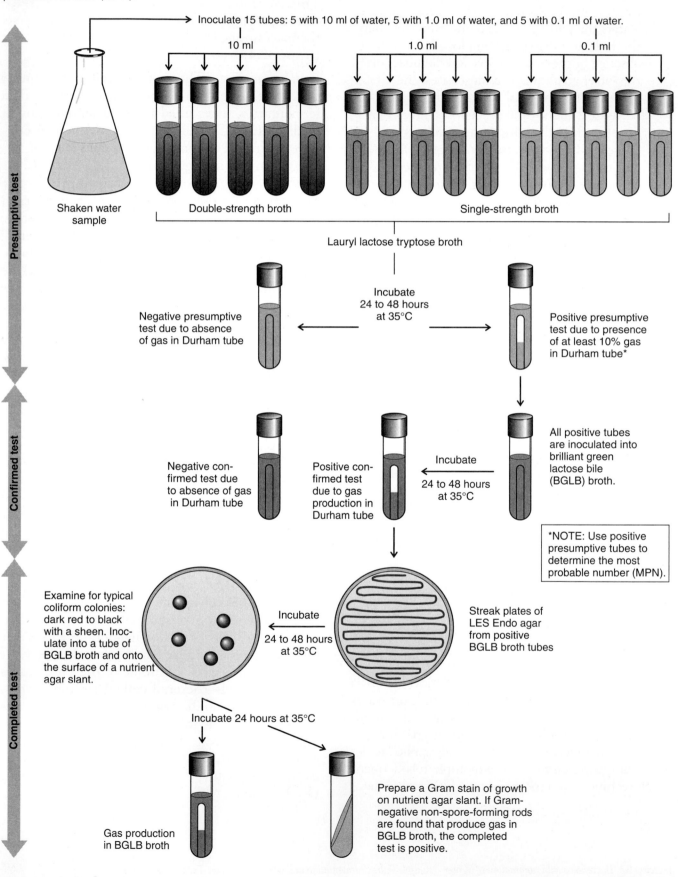

Inoculate 15 tubes: 5 with 10 ml of water, 5 with 1.0 ml of water, and 5 with 0.1 ml of water.

10 ml   1.0 ml   0.1 ml

Presumptive test

Shaken water sample

Double-strength broth

Single-strength broth

Lauryl lactose tryptose broth

Negative presumptive test due to absence of gas in Durham tube

Incubate 24 to 48 hours at 35°C

Positive presumptive test due to presence of at least 10% gas in Durham tube*

Confirmed test

Negative confirmed test due to absence of gas in Durham tube

Positive confirmed test due to gas production in Durham tube

Incubate 24 to 48 hours at 35°C

All positive tubes are inoculated into brilliant green lactose bile (BGLB) broth.

*NOTE: Use positive presumptive tubes to determine the most probable number (MPN).

Completed test

Examine for typical coliform colonies: dark red to black with a sheen. Inoculate into a tube of BGLB broth and onto the surface of a nutrient agar slant.

Incubate 24 to 48 hours at 35°C

Streak plates of LES Endo agar from positive BGLB broth tubes

Incubate 24 hours at 35°C

Gas production in BGLB broth

Prepare a Gram stain of growth on nutrient agar slant. If Gram-negative non-spore-forming rods are found that produce gas in BGLB broth, the completed test is positive.

Confirmed Test This test serves to confirm the presence of coliform bacteria when either a positive or doubtful presumptive test is obtained. A loopful of growth from such a presumptive tube is transferred into a tube of brilliant green bile 2% broth and incubated at 35°C for 48 hours. This is a selective medium for detecting coliform bacteria in water, dairy, and other food products. In order to do so, the correct concentration of the dye (brilliant green) and bile must be present. If the concentration is too great, coliform growth can also be inhibited. Bile is naturally found in the intestine, where it serves a similar purpose—encouraging growth of coliform bacteria while discouraging growth of other bacteria. A final selective agent in the medium is lactose. The broth tube also contains a Durham tube to detect gas production. The presence of gas in the Durham tube after incubating for 24–48 hours constitutes a positive confirmed test.

Completed Test This test helps to further confirm doubtful and, if desired, positive confirmed test results. The test has two parts:

1. A plate of LES Endo agar (or Levine's EMB agar) is streaked with a loopful of growth from a positive confirmed tube and incubated at 35°C for 18–24 hours. Typical coliform bacteria (*Escherichia coli* and *Enterobacter aerogenes*) exhibit good growth on this medium and form red to black colonies with a sheen. *Salmonella typhi* exhibits good growth, but the colonies are colorless. *Staphylococcus aureus* growth is inhibited altogether.

2. Next, a typical coliform colony from an LES Endo agar plate is inoculated into a tube of brilliant green bile broth and on the surface of a nutrient agar slant. They are then incubated at 35°C for 24 hours. A Gram stain is then prepared from growth present on the nutrient agar slant. The presence of gas in the brilliant green bile broth tube and Gram-negative, non-spore-forming rods constitutes a positive completed test for the presence of coliform bacteria; this, in turn, infers possible contamination of the water sample with fecal matter.

Membrane Filter Technique

For this technique, a known volume of water sample (100 ml) or of its dilutions is filtered by suction through a sterile polycarbonate or nitrocellulose acetate membrane filter. The filter is very thin (150 μm), and has a pore diameter of 0.45 μm. The precision manufacture of the filter is such that bacteria larger than 0.47 μm cannot pass through it. Filters with printed grid patterns are available for facilitating colony counting.

Once the water is filtered, the filter disc is aseptically transferred to the surface of a wetted pad contained in a petri dish. The pad is soaked with Endo broth MF on which coliform colonies appear to be pink to dark red in color and possess a golden green metallic sheen. After incubation at 35°C for 24 hours, the filter disc is examined for characteristic coliform colonies, and their number is determined. From this number, the total number of coliform bacteria present in the original water sample can be calculated. For additional confirmation, the procedure for the completed multiple-tube fermentation test can be applied.

The membrane filter method yields accurate results if the coliform colony count is in the range of 30–300 organisms per filter disc. Unknown samples require that various dilutions be examined. Turbidity of the sample constitutes a serious obstacle in cases where dilutions that yield coliform counts within the desired range contain sufficient suspended matter to clog the filter before the required volume of water has passed through.

The advantages of the membrane filter technique over the multiple-tube fermentation test are: (1) better reproducibility of results; (2) greater sensitivity, because larger amounts of water can be used; and (3) shorter time (one-quarter the time) for obtaining results. This method has been recognized by the United States Public Health Service for detection of coliforms in water.

Definitions

Aerobic bacteria. Microbes that grow and multiply in the presence of free gaseous oxygen.

Anaerobic bacteria. Microbes that grow best or exclusively in the absence of free oxygen.

Coliform bacteria. A collective term for bacteria that inhabit the colon, are Gram negative, and ferment lactose (see page 265).

Indicator organism. A nonpathogenic organism whose presence in water or sewage serves as a sign of possible pollution with pathogens.

Selective growth medium. A growth medium containing substances that inhibit the growth of certain organisms but not others.

Surfactant. A surface-active agent that forms a water-soluble common boundary between two substances. Examples include detergents and wetting agents.

Objectives

1. To introduce you to the use of a multiple-tube fermentation technique for detecting the presence and number of coliform indicator organisms present in water samples.
2. To introduce you to the use of an alternate method, the membrane filter technique, for detecting the presence and number of coliform bacteria in water samples.

References

McKinney, R. E. *Microbiology for sanitary engineers*, New York: McGraw-Hill Book Co., 1972.

Nester et al. *Microbiology: A human perspective*, 6th ed. The McGraw-Hill Companies, 2009. Chapter 31, Section 31.2.

Standard methods for the examination of water and wastewater, 18th ed. The American Water Works Association, 6666 West Quincy Ave., Denver, Co 80235.

Materials

Multiple-Tube Fermentation Technique (per student) (figure 32.4)

Test tubes, 50 ml, containing 10 ml of double-strength lauryl sulfate (lauryl lactose tryptose) broth plus Durham tubes, 5

Small test tubes containing 10 ml of single-strength lauryl sulfate broth plus Durham tubes, 10

Sterile 10-ml pipet, 1

Sterile 1-ml pipet calibrated in 0.1 ml units, 1

Brilliant green bile 2% broth plus Durham tubes, 2 tubes

LES Endo agar plate, 1

Nutrient agar slant, 1

Sterile 100-ml screw cap bottle for collecting water sample, 1

Membrane Filter Technique (demonstration) (figure 32.5)

A 1-liter side-arm Erlenmeyer flask, 1

Sterile membrane filter holder assembly, two parts wrapped separately (figure 32.5, frames 2 and 3), 1 unit

A metal clamp for clamping filter funnel to filter base

Sterile membrane filters, 47 mm diameter, 0.45 μm pore size

Forceps, 1 pair

Sterile 50-mm diameter petri dishes, 3

Absorbent filter pads, 3

Tube containing 10 ml of sterile Endo MF broth, 1

Sterile 90-ml water blanks, 2

Erlenmeyer flasks containing 25 ml of sterile water, 6

Sterile 10-ml pipets, 2

Vacuum pump or Venturi vacuum system

Water sample for coliform analysis, 100 ml

Procedure

Multiple-Tube Fermentation Technique

Note: Your instructor may ask you to bring a 50- to 100-ml sample of water from home, a nearby stream, a lake, or some other location for analysis. When taking a tap sample, the orifice of the water tap should be flamed before being opened. After opening, allow water to run for 5 to 10 minutes with the tap in the same position to prevent loosening of bacteria from inside the tap. Next, using aseptic technique, open a sterile bottle (obtained from the instructor beforehand) and collect a sample. If the sample cannot be examined within 1–2 hours, keep refrigerated until ready for use.

Figure 32.5 Analysis of water for fecal contamination. Cellulose acetate membrane filter technique. *(1)* Sterile membrane filter (0.45 μm) with grid for counting is handled with sterile forceps. *(2)* The sterile membrane filter is placed on filter holder base with grid side up. *(3)* The apparatus is assembled. *(4)* Sterile absorbent pads are aseptically placed in the bottom of three sterile petri dishes. *(5)* Each absorbent pad is saturated with 2.0 ml of Endo MF broth. *(6)* A portion of well-mixed water sample is poured into assembled funnel and filtered by vacuum. *(7)* Membrane filter is carefully removed with sterile forceps after disassembling the funnel. *(8)* Membrane filter is centered on the surface of the Endo-soaked absorbent pad (grid side up) using a slight rolling motion. *(9)* After incubation, the number of colonies on the filter is counted. The number of colonies on the filter reflects the number of coliform bacteria present in the original sample.

First Session (Presumptive Test)

1. Shake the water sample. Aseptically pipette 10-ml portions of the sample into each of the five large tubes containing 10-ml aliquots of double-strength lauryl sulfate broth. Next, with a 1-ml pipet, transfer 1-ml portions of the water sample into five of the smaller tubes and then 0.1-ml portions into the remaining five small tubes of lauryl sulfate broth. Be sure to label the tubes.
2. Incubate the test tubes for 48 hours at 37°C.

Second and Third Sessions (Presumptive and Confirmed Tests)

1. After 24 hours observe the tubes for gas production by gently shaking the tubes. If, after shaking, gas is not evident in the Durham tube, reincubate the tube for an additional 24 hours. Record any positive results for gas production in table 32.2 of the Laboratory Report.
2. Observe the tubes for gas production and turbidity after 48 hours of incubation. If neither gas nor turbidity is present in any of the tubes, the test is negative. If turbidity is present but gas is not, the test may be doubtful, since growth without gas may mean inhibition of coliform bacteria by noncoliform bacteria with shorter generation times. If at least 10% of the tube is filled with gas, the test is positive for coliform bacteria. Record your results in table 32.2 of the Laboratory Report.
3. MPN determination. Using your fermentation gas results in table 32.2, determine the number of tubes from each set containing 10% gas or more. Determine the MPN by consulting table 32.1; for example, if you have gas in two of the first five tubes, in two of the second five tubes, and none in the third three tubes, your test readout is 2-2-0. Table 32.1 shows that the MPN for this readout is 9. Thus, your water sample contains nine organisms per 100-ml water with a 95% statistical probability of there being between three and twenty-five organisms.

Note: If your readout for the series is 0-0-0, it means that the MPN is less than two organisms per 100 ml

of water. Also, if the readout is 5-5-5, it means the MPN is greater than 1,600 organisms/100 ml water. In this instance, what procedural modification is required to obtain a more significant result? Report your answer in question 10 of the Questions section in the Laboratory Report.

4. The confirmed test should be administered to all tubes demonstrating either a positive or doubtful presumptive test. Inoculate a loopful of growth from each tube showing gas or dense turbidity into a tube of brilliant green lactose bile 2% broth. Incubate the tube(s) at 37°C for 24–48 hours.

Note: For expediency, your instructor may instruct you to inoculate only one tube. If so, for the inoculum use the tube of lauryl sulfate broth testing positive with the least inoculum of water.

Fourth Session (Confirmed and Completed Tests)

1. Examine the brilliant green lactose bile 2% tube(s) for gas production. Record your findings in the confirmed test section of the Laboratory Report.
2. Streak a loopful of growth from a positive tube of brilliant green lactose bile 2% broth on the surface of a plate containing LES Endo agar. Incubate at 37°C for 24 hours.

Fifth Session (Completed Test)

1. Examine the LES Endo agar plate(s) for the presence of typical coliform colonies (dark red to black with a sheen). Record your findings in the completed test section of the Laboratory Report.
2. With a loop, streak a nutrient agar slant with growth obtained from a typical coliform colony found on the LES Endo agar plate. Also inoculate a tube of brilliant green lactose bile 2% broth with growth from the same colony. Incubate the tubes at 37°C for 24 hours.

Sixth Session (Completed Test)

1. Examine the brilliant green lactose bile 2% broth tube for gas production. Record your result in the completed test section of the Laboratory Report.

Table 32.1 MPN Index and 95% Confidence Limits for Various Combinations of Positive Results When Five Tubes Are Used Per Dilution (10 ml, 1.0 ml, 0.1 ml)

Combination of Positives	MPN Index/ 100 mL	95% Confidence Limits Lower	95% Confidence Limits Upper	Combination of Positives	MPN Index/ 100 mL	95% Confidence Limits Lower	95% Confidence Limits Upper
0-0-0	<2	—	—	4-3-0	27	12	67
0-0-1	3	1.0	10	4-3-1	33	15	77
0-1-0	3	1.0	10	4-4-0	34	16	80
0-2-0	4	1.0	13	5-0-0	23	9.0	86
1-0-0	2	1.0	11	5-0-1	30	10	110
1-0-1	4	1.0	15	5-0-2	40	20	140
1-1-0	4	1.0	15	5-1-0	30	10	120
1-1-1	6	2.0	18	5-1-1	50	10	150
1-2-0	6	2.0	18	5-1-2	60	30	180
2-0-0	4	1.0	17	5-2-0	50	20	170
2-0-1	7	2.0	20	5-2-1	70	30	210
2-1-0	7	2.0	21	5-2-2	90	40	250
2-1-1	9	3.0	24	5-3-0	80	30	250
2-2-0	9	3.0	25	5-3-1	110	40	300
2-3-0	12	5.0	29	5-3-2	140	60	360
3-0-0	8	3.0	24	5-3-3	170	80	410
3-0-1	11	4.0	29	5-4-0	130	50	390
3-1-0	11	4.0	29	5-4-1	170	70	480
3-1-1	14	6.0	35	5-4-2	220	100	580
3-2-0	14	6.0	35	5-4-3	280	120	690
3-2-3	17	7.0	40	5-4-4	350	160	820
4-0-0	13	5.0	38	5-5-0	240	100	940
4-0-1	17	7.0	45	5-5-1	300	100	1300
4-1-0	17	7.0	46	5-5-2	500	200	2000
4-1-1	21	9.0	55	5-5-3	900	300	2900
4-1-2	26	12	63	5-5-4	1600	600	5300
4-2-0	22	9.0	56	5-5-5	≥1600	—	—
4-2-1	26	12	65				

From *Standard Methods for the Examination of Water and Wastewater,* 18th edition. Copyright 1992 by the American Public Health Association.

2. Prepare a Gram stain of some of the growth present on the nutrient agar slant. Examine the slide for the presence of Gram-negative, non-spore-forming rods. Record your results in the completed test section of the Laboratory Report. The presence of gas and of Gram-negative, non-spore-forming rods constitutes a positive completed coliform test.

Membrane Filter Technique

First Session

1. Shake the water sample. Prepare two dilutions by transferring successive 10-ml aliquots into 90-ml blanks of sterile water (10^{-1} and 10^{-2} dilutions).

   Note: Reshake prepared dilutions before using.

2. Assemble the filter holder apparatus as follows (figure 32.5):
   a. Using aseptic technique, unwrap the lower portion of the filter base, and insert a rubber stopper.
   b. Insert the base in the neck of the 1-liter side-arm Erlenmeyer flask (figure 32.5, frame 2).
   c. With sterile forceps (sterilize by dipping in alcohol and flaming them with the Bunsen burner), transfer a sterile membrane filter onto the glass or plastic surface of the filter holder base (figure 32.5, frames 1 and 2). Make certain the membrane filter is placed with the ruled side up.
   d. Aseptically remove the covered filter funnel from the butcher paper, and place the lower surface on top of the membrane filter. Clamp the filter funnel to the filter base with the clamp provided with the filter holder assembly (figure 32.5, frame 3).
3. Prepare 3 plates of Endo MF medium by adding 2-ml aliquots of the tubed broth to sterile absorbent pads previously placed aseptically (using sterile tweezers) on the bottom of the three petri dishes (figure 32.5, frames 4 and 5).
4. Remove the aluminum filter cover and pour the highest water dilution (10^{-2}) into the funnel (figure 32.5, frame 6). Assist the filtration process by turning on the vacuum pump or Venturi vacuum system.
5. Rinse the funnel walls with two 25-ml aliquots of sterile water.
6. Turn off (break) the vacuum and remove the filter holder funnel. Using aseptic technique and sterile tweezers, transfer the filter immediately to the previously prepared petri dish (figure 32.5, frames 7 and 8). Using a slight rolling motion, center the filter, grid side up, on the medium-soaked absorbent pad. Take care not to trap air under the filter, as this will prevent nutrient media from reaching all of the membrane surface (figure 32.5, frame 9).
7. Reassemble the filter apparatus with a new membrane filter, and repeat the filtration process first with the 10^{-1} water sample and finally with a 100-ml aliquot of the undiluted water sample.
8. Label and invert the petri dishes to prevent any condensate from falling on the filter surface during incubation. Incubate plates for 24 hours at 37°C.

Second Session

1. Count the number of coliform bacteria by using either the low-power objective lens of the microscope or a dissecting microscope. Count only those colonies that exhibit a pink to dark-red center with or without a distinct golden green metallic sheen.
2. Record the number of colonies found in each of the three dilutions in table 32.3 of the Laboratory Report.

EXERCISE

32

Laboratory Report: Bacteriological Examination
of Water: Multiple-Tube Fermentation
and Membrane Filter Techniques

Results

1. Multiple-Tube Fermentation Technique
   a. Record the results of the presumptive test in table 32.2.

Table 32.2   Presumptive Test for the Presence or Absence of Gas and Turbidity in Multiple-Tube Fermentation Media

Water Sample Size (ml)	Presence of Gas and Turbidity*				
	Tube #1	Tube #2	Tube #3	Tube #4	Tube #5
10					
1					
0.1					

*Use a (+) sign to indicate gas and a circle (O) around the plus sign to indicate turbidity.

   b. Determine the MPN below:

   Test readout: _____   MPN: _____   95% Confidence Limits: _____

   c. Confirmed test results (gas production in brilliant green lactose bile 2% broth):

   Sample Number   **Gas (+ or −)**
                       24 hours   48 hours

   d. Appearance of colonies on LES Endo agar:

   e. Completed test results:

   Sample Number   **Gas (+ or −)**   **Gram Stain Reaction**

2. Membrane filter technique

Table 32.3   Number of Coliform Colonies Present in Various Dilutions of the Water Sample

Undiluted Sample	10⁻¹ Dilution	10⁻² Dilution

a. Calculate the number of coliform colonies/ml present in the original water sample (show your calculations):

Questions

1. What nutritional means could be used to speed up the growth of the coliform organisms when using the membrane filter technique?

2. Describe two other applications of the membrane filter technique.

3. Why not test for pathogens such as *Salmonella* directly rather than using an indicator organism, such as the coliform bacteria?

4. Why does a positive presumptive test not necessarily indicate that the water is unsafe for drinking?

5. List three organisms that are apt to give a positive presumptive test.

6. Describe the purpose of lactose and LES Endo agar in these tests.

7. What are some limitations of the membrane filter technique?

8. Define the term *coliform*.

9. Briefly explain what is meant by presumptive, confirmed, and completed tests in water analysis.

10. See Note 3, MPN determination, on page 270 for the question.

INTRODUCTION to Biotechnology

Watson and Crick first proposed the structure of DNA in the early 1950s. In less than 50 years, the field had mushroomed: geneticists had isolated DNA, transferred specific genes to another organism, and determined the sequence of bases in the DNA of specific genes as well as entire bacterial genomes. These sequences have been used to identify or classify an organism and to determine evolutionary relationships.

The techniques for manipulating DNA have many applications. A particular gene may be cut from the DNA of one organism using restriction enzymes and inserted into another organism, so that the action of the gene can be studied independent of the organism. DNA manipulation is also a method for obtaining a gene product, usually a protein, in large quantities. For instance, the human gene for insulin can be cloned into yeast or bacterial DNA. These microorganisms then can be grown in huge quantities, and the insulin can be purified for use in the treatment of diabetes.

In these next two exercises, you will apply some of the techniques used in biotechnology to identify DNA. First you will use restriction enzymes to produce patterns of DNA unique to each organism. In the second exercise, you will use a computer database to identify an organism from its DNA sequences of the 16S rRNA gene.

NOTES:

EXERCISE

33

Identifying DNA with Restriction Enzymes

Getting Started

DNA from one organism can be distinguished from the DNA from another organism using restriction enzymes. Restriction enzymes not only cut DNA at very specific sequences, but also there are many different enzymes, each with its own sequence. For example, the enzyme Hha I cuts at the sequence GCGC. In the two sequences of DNA shown below, each strand is cut in two pieces because each contains GCGC; the pieces, however, differ in size between the two sequences.

```
ATGGCTCAA GCGC TCACGGTAACTGCTGCATC-CCGTTATACGAGCTACT
TACCGAGTT CGCG AGT-GCCATTGACGACGTAGGGCAATATGCTCGATGA
```

```
ATATCGTTGAACTCCGTGTAGACT GCGC ACGTG-TTACAATCCACCAAGT
TATAGCAACTTGAGGCACATCTGA CGCG TGCACAATGTTAGGTGGTTCA
```

The location of the specific sequences varies from species to species, and only identical strands of DNA are cut into the same number and size of fragments. This is the basis for comparing DNA from different organisms, one that has many applications. For example, in forensic (legal) investigations, the DNA from bloodstains can be compared with the DNA of a suspected murderer. In epidemiology investigations, the DNA from a serious *E. coli* outbreak of diarrhea can be compared to the DNA isolated from other *E. coli* strains found in food to determine the source of infection.

Where do these enzymes come from? They are found in bacteria, which use them to degrade foreign DNA that might enter their cell. Each cell methylates its DNA by adding a methyl group at a particular site and thereby prevents its own DNA from being degraded by its own restriction enzymes. Foreign DNA entering the cell does not have this specific pattern of methylation, and so the cell cleaves it, restricting its expression in the cell. Therefore, these enzymes are called **restriction enzymes.** They can be isolated from bacteria and used in the laboratory for studying and manipulating

DNA. The restriction enzymes' names are derived from the first letter of the bacteria's genus and the first two letters of their species. EcoRI is the first restriction enzyme from *Escherichia coli* strain R, for example.

In the laboratory, samples of the DNA to be compared are mixed with a restriction enzyme and incubated until the enzymes have cleaved the DNA at the recognition site unique to the enzyme. Each sample of DNA contains pieces of DNA of specific lengths. How, then, do you determine the size of the fragments? This can be done by running the DNA on an **electrophoresis** agarose gel.

Electrophoresis gel is made from a highly refined agar, called agarose (figure 33.1). Holes or wells are formed in the agarose when it is poured in the mold. The gel is covered with buffer, and the DNA that has been cut is placed in the wells. A current is applied, and the negatively charged DNA moves to the positive pole. The smaller the pieces, the more

---

Figure 33.1 Gel electrophoresis of three DNA samples, including a size standard.

quickly they move through the porous agarose gel. Since we cannot see the DNA in the gel, dyes are added to the sample; these travel through the agar at about the same speed as the DNA, providing an idea of how far the bands have traveled. If the gel runs too long, the samples run off the bottom of the gel and diffuse into the **buffer solution.**

Ethidium bromide can be added to the gel before it is poured. This is a dye that stains DNA, and when the gel is placed on a UV transilluminator after electrophoresis, the fragments of DNA can be seen and compared with each other. This pattern is sometimes called a DNA fingerprint.

In this exercise, you will compare the DNA of three bacterial viruses: phage lambda, phage ϕX 174, and virus X (which is either lambda or ϕX 174). The restriction enzyme Dra I is used to cut the DNA. It recognizes the base sequences

TTTAAA

AAATTT

Lambda contains thirteen sites that can be cut by this enzyme, but ϕX 174 has only two. Therefore, we should be able to identify virus X as either ϕX 174 or lambda, depending on the number and size of the pieces separated by gel electrophoresis.

Note: Students should practice using micropipettors and loading a practice gel before doing the exercise.

Definitions

Buffer solution. A salt solution formulated to maintain a particular pH.

Electrophoresis. A procedure used to separate components by electrical charge.

Restriction enzymes. Enzymes isolated from bacteria that are used to cut DNA at specific sequences.

Objectives

1. To understand the use of restriction enzymes to cut DNA into segments.
2. To learn how to separate different lengths of DNA using gel electrophoresis.
3. To understand the use of DNA fingerprinting to identify DNA.

References

Alcamo, E. I. *DNA technology: The awesome skill,* 2nd ed. San Diego: Harcourt Academic Press, 2001.

Nester et al. *Microbiology: A human perspective,* 6th ed. The McGraw-Hill Companies, 2009. Chapter 9, Section 9.1.

Sambrook, J.; Fritsch, E. F.; and Maniatis, T. *Molecular cloning,* 2nd ed. Cold Spring Harbor Laboratory Press, 1989.

Materials

Lambda DNA

ϕX 174 DNA

Unknown phage (either lambda or ϕX 174)

DNA size standard

Restriction enzyme Dra I

Microfuge tubes or Eppendorf tubes

Micropipettors

Gel box

Agarose

TBE buffer for electrophoresis (TRIS-EDTA-Borate)

TE buffer

Ethidium bromide

Stop mix (includes tracking dye)

Water bath

Microfuge

UV transilluminator

Goggles

Spatula

Disposable gloves

Distilled water

Procedure

Warning: Ethidium bromide is a potent mutagen. Always wear gloves when handling.

1. Pour an agarose gel into a gelbox. The amount of agar will depend on the size of the gel box. The usual concentration is 0.7% agar, and it is

dissolved in the buffer TBE. Add ethidium bromide at 15 µl/200 ml agarose gel.

2. Add sterile distilled water (as determined by the instructor) to a sterile Eppendorf tube, and then add 1µl 10X TE buffer. Buffer is added before the enzyme so that the conditions are immediately optimal for the enzyme.

3. Add lambda DNA to the tube. The amount added is usually 0.5–1.0 µg/µml. The instructor will indicate how much you should add.

4. Add the restriction enzyme Dra I.

5. Repeat for ϕX 174 DNA and the unknown virus DNA.

6. Each tube contains:
   Distilled water to bring total volume to 10 µl, including enzyme.
   10X TE buffer
   DNA
   Enzyme

7. Mix and incubate for 30 minutes in a 37°C water bath.

8. Add stop mix (usually 1 µl of 10X). This not only contains dyes, but also stops the reaction and helps the sample sink to the bottom of the well.

9. Place gel in a gel box. Cover the agarose gel with running buffer, and load the samples into the wells. Include a size standard in one of the lanes. This is DNA of known length so that you can estimate the size of the cut DNA pieces of the phage.

10. Attach the electrodes to the gel. Attach the positive electrode at the bottom of the gel. DNA is negatively charged, so it will migrate to the positive electrode. (Remember, run to red.) Run the gel until the colored bands of the dyes are separated. These dyes travel at about the same speed as the DNA and indicate how far the DNA has migrated.

11. Transfer the gel to a transilluminator with a spatula while wearing gloves.

12. View on a UV transilluminator.
    Warning: It is essential to wear goggles so that the UV rays do not damage your eyes.

13. Photograph the gel or make a drawing of the bands. Compare the number of bands that result from cutting of the phage DNA with Dra I.

NOTES:

EXERCISE

33

Laboratory Report:
Identifying DNA with Restriction Enzymes

Results

1. Record your results in table 33.1.

Table 33.1   Results of Phage DNA Electrophoresis

	Lambda	ϕX 174	Unknown (Phage X)
Number of fragments			
About what size is the largest piece?			
About what size is the smallest piece?			

2. What is the probable identity of phage X?

Questions

1. How were you able to estimate the size of the DNA fragments?

2. What is the purpose of adding ethidium bromide to the agarose gel?

3. What is the purpose of adding a tracking dye to the DNA before adding to the well?

4. Would you have the same number of fragments from each phage if you used a different restriction enzyme? Why or why not?

EXERCISE

34

Identification of Bacteria
Using the Ribosomal Data Project

Getting Started

Identifying and classifying bacteria have always been more difficult than identifying and classifying plants and animals. Bacteria have very few differences in their structure, and while they are metabolically extremely diverse, it has not been clear which of these characteristics are the most important for identification or grouping. For example, is nitrogen fixation more significant than anaerobic growth, or endospore formation more important than photosynthesis?

The ability to sequence the DNA of microorganisms has offered a new solution. The closer organisms are related to each other, the more similar their **DNA nucleotide sequences.** Certain genes common to all bacteria can be sequenced and compared. What genes should be compared? The DNA coding for a part of the ribosome, namely the **16S** ribosome portion, is a good choice. Ribosomes are critically important for protein synthesis, and any mutation is quite likely to be harmful. Some mutations, however, are neutral or perhaps even advantageous, and these mutations then become part of the permanent genome. These sequences change very slowly over time and are described as highly conserved. The 16S rRNA DNA segment is found in all organisms (slightly larger in eukaryotes) with the same function, so the sequences can easily be compared.

This approach has had exciting results. Some specific base sequences are always found in some organisms and not others. These are called **signature sequences.** Also, new relationships among bacteria can be determined by comparing sequences of organisms base by base by means of computer programs. The more the sequences diverge, the more the organisms have evolved from one another.

Perhaps even more useful, these sequences can be used to identify bacteria. The sequences of at least 16,000 organisms are in public databases. If a new, unknown sequence is submitted to a database management computer, it will respond in seconds with the most likely identification of the species containing the sequence.

This exercise will give you a chance to send the DNA sequence of the 16S ribosomal RNA of an organism to the Ribosomal Database Project (RDP) and to identify the organism. Since many schools do not have the resources for determining the specific sequences, you will be given the sequences, which you can enter on the Web and immediately receive an identification. You may also look for other sequences posted on the Web or in microbiological journals, such as *Journal of Bacteriology,* published by the American Society for Microbiology.

Definitions

DNA nucleotide sequence. The order that bases are found in a piece of DNA.

16S. S is an abbreviation for Svedberg. It is a unit of mass that is measured by the rate a particle sediments in a centrifuge. The prokaryotic ribosome is made up of two main parts: 30S and 50S. The 30S particle is made up of the 16S rRNA plus twenty-one polypeptide chains.

Signature sequences. DNA sequences of about 5–10 bases long found at a particular location in the 16S rRNA. These DNA sequences are unique to Archaea, Bacteria, or Eukarya (eukaryotes).

Objectives

1. To understand the importance of the 16S rRNA sequence for the identification of organisms.
2. To understand how to identify organisms using the Ribosomal Database Project.

Reference

Nester et al. *Microbiology: A human perspective,*
6th ed. The McGraw-Hill Companies, 2009.
Chapter 10, Section 10.3.

Procedure

1. Open your internet browser (Netscape or Explorer, for example) on the computer.
2. Type in the URL http://rdp.cme.msu.edu
3. Under "RDP Analysis Tools," click "Sequence Match."
4. Type the sequence in the box labeled "Cut and Paste Sequence(s)." This is easier to do if one person reads the sequence while the other person types in the letters. Sometimes only a few hundred bases are necessary for an identification.
5. Click "Submit."
6. Read identification under "Hierarchy View."

Hint: The organisms were studied in:

1. Exercise 31
2. Exercise 22
3. Exercise 23
4. Exercise 13 (genus only)
5. Photosynthetic bacteria (not found in any exercise)

Organism 1

```
   1   tctctgatgt tagcggcgga cgggtgagta acacgtggat aacctaccta taagactggg
  61   ataacttcgg gaaaccggag ctaataccgg ataatatttt gaaccgcatg gttcaaaagt
 121   gaaagacggt cttgctgtca cttatagatg gatccgcgct gcattagcta gttggtaagg
 181   taacggctta ccaaggcaac gatgcatagc cgacctgaga gggtgatcgg ccacactgga
 241   actgagacac ggtccagact cctacgggag gcagcagtag ggaatcttcc gcaatgggcg
 301   aaagcctgac ggagcaacgc cgcgtgagtg atgaaggtct tcggatcgta aaactctgtt
 361   attagggaag aacatatgtg taagtaactg tgcacatctt gacggtacct aatcagaaag
 421   ccacggctaa ctacgtgcca gcagccgcgg taatacgtag gtggcaagcg ttatccggaa
 481   ttattgggcg taaagcgcgc gtaggcggtt ttttaagtct gatgtgaaag cccacggctc
 541   aaccgtggag ggtcattgga aactggaaaa cttgagtgca gaagaggaaa gtggaattcc
 601   atgtgtagcg gttaaatgcg cagagatatg gaggaacacc agtggcgaag cgactttct
 661   ggtctgtaac tgacgctgat gtgcgaaagc gtgggaatca acaggatta gataccctgg
 721   tagtccacgc cgtaaacgat gagtgctaag tgttagggggg tttccgcccc ttagtgctgc
 781   agctaacgca ttaagcactc cgcctgggga gtacgaccgc aaggttgaaa ctcaaaggaa
 841   ttgacgggga cccgcacaag cggtggagca tgtggtttaa ttcgaagcaa cgcgaagaac
 901   cttaccaaat cttgacatcc tttgacaact ctagagatag agccttcccc ttcgggggac
 961   aaagtgacag gtggtgcatg gttgtcgtca gctcgtgtcg tgagatgttg ggttaagtcc
1021   cgcaacgagc gcaacccta agcttagttg ccatcattaa gttgggcact ctaagttgac
1081   tgccggtgac aaaccggagg aaggtgggga tgacgtcaaa tcatcatgcc ccttatgatt
1141   tgggctacac acgtgctaca atggacaata caaagggcag cgaaaccgcg aggtcaagca
1201   aatcccataa agttgttctc agttcggatt gtagtctgca actcgactac atgaagctgg
1261   aatcgctagt aatcgtagat cagcatgcta cggtgaatac gttcccgggt cttgtacaca
1321   ccgcccgtca caccacgaga gtttgtaaca
```

Organism 2

```
  1    taacacgtgg ataacctacc tataagactg ggataacttc gggaaaccgg agctaatacc
 61    ggataatata ttgaaccgca tggttcaata gtgaaagacg gttttgctgt cacttataga
121    tggatccgcg ccgcattagc tagttggtaa ggtaacggct taccaaggca acgatgcgta
181    gccgacctga gagggtgatc ggccacactg gaactgagac acggtccaga ctcctacggg
241    aggcagcagt agggaatctt ccgcaatggg cgaaagcctg acggagcaac gccgcgtgag
301    tgatgaaggt cttcggatcg taaaactctg ttattaggga agaacaaatg tgtaagtaac
361    tatgcacgtc ttgacggtac ctaatcagaa agccacggct aactacgtgc
```

Organism 3

```
   1    gcctaataca tgcaagtaga acgctgagaa ctggtgcttg caccggttca aggagttgcg
  61    aacgggtgag taacgcgtag gtaacctacc tcatagcggg ggataactat tggaaacgat
 121    agctaatacc gcataagaga gactaacgca tgttagtaat ttaaaagggg caattgctcc
 181    actatgagat ggacctgcgt tgtattagct agttggtgag gtaaaggctc accaaggcga
 241    cgatacatag ccgacctgag agggtgatcg gccacactgg gactgagaca cggcccagac
 301    tcctacggga ggcagcagta gggaatcttc ggcaatgggg caaccctga ccgagcaacg
 361    ccgcgtgagt gaagaaggtt ttcggatcgt aaagctctgt tgttagagaa gaatgatggt
 421    gggagtggaa aatccaccaa gtgacggtaa ctaaccagaa agggacggct aactacgtgc
 481    cagcagccgc ggtaatacgt aggtcccgag cgttgtccgg atttattggg cgtaaagcga
 541    gcgcaggcgg tttttttaagt ctgaagttaa aggcattggc tcaaccaatg tacgctttgg
 601    aaactggaga acttgagtgc agaaggggag agtggaattc catgtgtagc ggtgaaatgc
 661    gtagatatat ggaggaacac cggtggcgaa agcggctctc tggtctgtaa ctgacgctga
 721    ggctcgaaag cgtggggagc aaagaggatt agataccctg gtagtccacg ccgtaaacga
 781    tgagtgctag gtgttaggcc ctttccgggg cttagtccg gagctaacgc attaagcact
 841    ccgcctgggg agtacgaccg caaggttgaa actcaaagga attgacgggg gcccgcacaa
 901    gcggtggagc atgtggttta attcgaagca acgcgaagaa ccttaccagg tcttgacatc
 961    ccgatgcccg ctctagagat agagttttac ttcggtacat cggtgacagg tggtgcatgg
1021    ttgtcgtcag ctcgtgtcgt gagatgttgg gttaagtccc gcaacgagcg caaccccctat
1081    tgttagttgc catcattaag ttgggcactc tag
```

Organism 4

```
  1    ggtaccactc ggcccgaccg aacgcactcg cgcggatgac cggccgacct ccgcctacgc
 61    aatacgctgt ggcgtgtgtc cctggtgtgg gccgccatca cgaagcgctg ctggttcgac
121    ggtgtttat gtaccccacc actcggatga gatgcgaacg acgtgaggtg gctcggtgca
181    cccgacgcca ctgattgacg cccccctcgtc ccgttcggac ggaacccgac tgggttcagt
241    ccgatgccct taagtacaac agggtacttc ggtggaatgc gaacgacaat ggggccgccc
301    ggttacacgg gtggccgacg catgactccg ctgatcggtt cggcgttcgg ccgaactcga
361    ttcgatgccc ttaagtaata acgggtgttc cgatgagatg cgaacgacaa tgaggctatc
421    cggcttcgtc cgggtggctg atgcatctct tcgacgctct ccatggtgtc ggtctcactc
481    tcagtgagtg tgattcgatg cccttaagta ataacgggcg ttacgaggaa ttgcgaacga
541    caatgtggct acctggttct cccaggtggt taacgcgtgt tcctcgccgc cctggtgggc
601    aaacgtcacg ctcgattcga gcgtgattcg atgcccttaa gtaataacgg ggcgttcggg
661    gaaatgcgaa cgtcgtcttg gactgatcgg agtccgatgg gtttatgacc tgtcgaactc
721    tacggtctgg tccgaaggaa tgaggattcc acacctgcgg tccgccgtaa agatggaatc
781    tgatgttagc cttgatggtt tggtgacatc caactggcca cgacgatacg tcgtgtgcta
841    agggacacat tacgtgtccc cgccaaacca agacttgata gtcttggtcg ctgggaacca
901    tcccagcaaa ttccggttga tcctgccgga ggccattgc
```

Organism 5

```
   1  agagtttgat cctggctcag agcgaacgct ggcggcaggc ttaacacatg caagtcgaac
  61  gggcgtagca atacgtcagt ggcagacggg tgagtaacgc gtgggaacgt accttttggt
 121  tcggaacaac acagggaaac ttgtgctaat accggataag cccttacggg gaaagattta
 181  tcgccgaaag atcggcccgc gtctgattag ctagttggtg aggtaatggc tcaccaaggc
 241  gacgatcagt agctggtctg agaggatgat cagccacatt gggactgaga cacggcccaa
 301  actcctacgg gaggcagcag tggggaatat tggacaatgg gcgaaagcct gatccagcca
 361  tgccgcgtga gtgatgaagg ccctagggtt gtaaagctct tttgtgcggg aagataatga
 421  cggtaccgca agaataagcc ccggctaact tcgtgccagc agccgcggta atacgaaggg
 481  ggctagcgtt gctcggaatc actgggcgta aagggtgcgt aggcgggttt ctaagtcaga
 541  ggtgaaagcc tggagctcaa ctccagaact gcctttgata ctggaagtct tgagtatggc
 601  agaggtgagt ggaactgcga gtgtagaggt gaaattcgta gatattcgca agaacaccag
 661  tggcgaaggc ggctcactgg gccattactg acgctgaggc acgaaagcgt ggggagcaaa
 721  caggattaga taccctggta gtccacgccg taaacgatga atgccagccg ttagtgggtt
 781  tactcactag tggcgcagct aacgctttaa gcattccgcc tggggagtac ggtcgcaaga
 841  ttaaaactca aaggaattga cgggggcccg cacaagcggt ggagcatgtg gtttaattcg
 901  acgcaacgcg cagaacctta ccagcccttg acatgtccag gaccggtcgc agagacgtga
 961  ccttctcttc ggagcctgga gcacaggtgc tgcatggctg tcgtcagctc gtgtcgtgag
1021  atgttgggtt aagtcccgca acgagcgcaa ccccgtcct tagttgctac catttagttg
1081  agcactctaa ggagactgcc ggtgataagc cgcgaggaag gtggggatga cgtcaagtcc
1141  tcatggccct tacgggctgg gctacacacg tgctacaatg gcggtgacaa tgggaagcta
1201  aggggtgacc cttcgcaaat ctcaaaaagc cgtctcagtt cggattgggc tctgcaactc
1261  gagcccatga agttggaatc gctagtaatc gtggatcagc atgccacggt gaatacgttc
1321  ccgggccttg tacacaccgc ccgtcacacc atgggagttg gctttacctg aagacggtgc
1381  gctaaccagc aatgggggca gccggccacg gtagggtcag cgactggggt gaagtcgtaa
1441  caaggtagcc gtaggggaac ctgcggctgg atcacctcct t
```

EXERCISE

34

Laboratory Report: Identification of Bacteria
Using the Ribosomal Data Project

Results

Organism	Bacteria or Archaea
1 _____	_____
2 _____	_____
3 _____	_____
4 _____	_____
5 _____	_____

Questions

1. What is an advantage of identifying an organism using the Ribosomal Data Project?

2. What is a disadvantage?

3. Microbiologists in general think that you cannot randomly create a sequence that the database will identify as an organism. Can you prove them wrong?

INTRODUCTION   to the Individual Projects

Many students taking a laboratory class in microbiology are planning careers in medicine and health-related fields. Therefore, the emphasis in many courses is on laboratory exercises that will provide vital skills and concepts to prepare the students for those occupations. The laboratory exercises tend to emphasize the control and identification of pathogenic organisms in addition to an understanding of basic microbiological principles.

Pathogenic organisms, however, make up a very small percentage of the known microorganisms that exist in the world. For the most part, it is the nonpathogenic prokaryotes that are responsible for recycling animal and plant material and are absolutely essential for making life on earth possible.

How are they able to carry out these activities? Prokaryotes have astonishing versatility. The physiological abilities of prokaryotes make the eukaryotes appear very limited. For instance, no eukaryotes can fix nitrogen, oxidize sulfur for energy, or produce methane, and very few can grow anaerobically. The unique physiological tricks that bacteria use in breaking down and synthesizing molecules as well as their adaptation to their niche are very interesting. For instance, there are organisms that specialize in taking one-carbon compounds produced by plants and building them into cellular material that can then be utilized by other organisms. Others break down complex molecules to simpler components to be incorporated by other bacteria. Some bacteria have such adaptive strategies as producing light or resisting the effects of radiation,

which are clearly important to the organisms' survival, although their roles are not understood.

Once you sample the world of microbiology, you may find it especially exciting, and perhaps you would like to investigate an organism on your own as an individual project. The following section presents some protocols for isolating bacteria that are particularly interesting for their unique physiological abilities. Some of these organisms require patience and persistence to isolate. A large part of the project is finding the necessary equipment and source of samples. But if you enjoy a challenge, these individual projects can be very rewarding. Your instructor will decide whether the results should be presented in a written paper, a poster, a report to the class, or both a written and an oral presentation.

Isolating these organisms is also a true investigation. The best source of some of these organisms has not yet been determined. It would be very helpful for future students if the results of these independent investigations were filed in the laboratory, in a separate folder for each organism. You could then add information to the folder about what you learned from your experience with isolating and identifying your particular organism.

Two projects are included in the physical manual on the following pages (Hydrocarbon-Degrading Bacteria and Luminescent Bacteria). Two more are posted online at www.mhhe.com/Nester6e (Methylotrophs, and The UV-Resistant *Deinococcus*).

NOTES:

INDIVIDUAL PROJECT 35

Hydrocarbon-Degrading Bacteria: Cleaning Up After Oil Spills

Getting Started

The biosphere contains a great variety of organisms that are collectively capable of breaking down just about all naturally occurring carbon compounds. This is fortunate, because otherwise any compound not reduced to simpler molecules would accumulate in the environment. Sometimes it is useful to isolate the specific bacteria that are responsible for breaking down a particular compound. This can be done by using a medium containing only that particular compound as the source of carbon. If you want to isolate an organism, for instance, that oxidizes phenol, you can use a mineral salts medium for the nitrogen, phosphate, and sulfur and then use phenol for the carbon source.

Crude petroleum oil spills are a serious threat to the marine environment. Several methods of removing the oil from the environment were tried in a major oil spill in Alaska. One of the more useful was simply adding fertilizer. It provided a source of nitrogen and phosphorus to encourage the growth of bacteria already present in the environment.

The hydrocarbons in oil are natural compounds found almost everywhere in nature. They are made up of a mix of different lengths of carbon chains saturated (covered) with hydrogen, called alkanes, and other carbon-hydrogen compounds. Since petroleum is formed by fossil plants, these same organic molecules are found in garden soil produced by growing plants. The same kinds of organisms that are present in marine environments breaking down seaweed (and potentially petroleum) are also present in the garden soil. Therefore, even though muddy marine soil is an ideal source for these organisms, garden soil is also an excellent source of hydrocarbon degraders.

Choose any petroleum product—kerosene or fuel oil—for example for the carbon source. Avoid products containing detergents or other additives.

Enrichment

The first step in these kinds of isolations is enrichment, a way to increase the numbers of the desired bacteria. This means adding inoculum consisting of a source of the organism to a mineral medium with the sole carbon source. In this case, soil will be added to the mineral medium, plus a hydrocarbon. The soil itself contains nutrients that can support the growth of many organisms, but the bacteria that are able to also use the hydrocarbon will have an advantage. They will be able to grow after the carbon compounds added with the soil have been exhausted.

1. For the enrichment, prepare mineral salts medium (MSM) broth (table 35.1). Add 200 ml to a 500-ml flask and cover with a foil cap. Prepare several flasks to maximize your chances of successfully isolating the organisms.

Table 35.1 Minimal Salts Medium (Modified from E. Rosenberg)

NaCl	2.5 g (28.4g NaCl for marine organisms)
K_2HPO_4	4.74 g
KH_2PO_4	0.56 g
$MgSO_4 \cdot 7H_2O$	0.50 g
$CaCl_2 \cdot H_2O$	0.1 g
NH_4NO_3	2.5 g
Tap water	1 liter
Agar (for plates)	20 grams (15 grams Difco agar)
pH 7.1	

Note: 2.06 $(NH_4)_2SO_4$ and 3.15 KNO_3 can be substituted for 2.5 g NH_4NO_3

2. Add 0.1% vol/vol of the hydrocarbon of the carbon source (0.2 ml hydrocarbon/200 ml broth). Remember: %vol/vol = ml/100 ml.
3. Add about a gram of soil to the broth.

4. Incubate the flask for 2 weeks at room temperature. These organisms are usually obligate aerobes, so shaking the flask is helpful, but not necessary. Avoid light to discourage the growth of algae. Since they fix CO_2, the algae can grow on CO_2 from the air without degrading the hydrocarbon.

Selection

The second step is selection. After enriching for the organism in a broth culture, your chances of isolating it are greatly increased. Streak the broth on mineral salts agar plates and add the hydrocarbon as described. The medium is very selective because only organisms able to utilize the hydrocarbon can grow.

1. Pour mineral salts agar plates. If you use plastic petri plates, check to see if the hydrocarbon you plan to use dissolves them. If it does, use glass plates.
2. Add the hydrocarbon carbon source, such as fuel oil. (Since some petroleum products do not dissolve well in water, it was not placed in the agar.) Invert the agar plate, and place a piece of filter paper in the lid of the petri dish. Add 0.5 ml of the hydrocarbon to the filter paper, and replace the inverted top of the agar plate (figure 35.1). The organism will be able to grow on the fumes.
3. To obtain isolated colonies, streak the MSM agar plates with a loopful of inoculum from your enrichment flasks.
4. Prepare a control by adding water to the filter paper in another plate. This will control for organisms that might be able to degrade agar

to obtain carbon and therefore would grow using agar instead of the hydrocarbon.
5. Incubate the plates at room temperature for about a week.

Isolation

1. Examine your plates carefully. Is there more than one type of colony? Note if any organisms are growing on the control plate—these must be using either nutrients contaminating the minerals in the medium or the agar itself (a few bacteria can utilize agar as a carbon source).
2. Purify your isolates by restreaking the organisms on a similar plate with the same hydrocarbon. Repeat until you can be sure you have at least one pure culture.
3. Make a Gram stain for an initial identification of the organism or organisms growing on the plates. If a Gram-negative rod is present, determine if it is oxidase positive. *Pseudomonas* is oxidase positive and well known for its ability to degrade unusual compounds. There are many Gram-positive organisms as well as other Gram-negative bacteria that are also involved in degrading complex molecules. It may not be possible for you to specifically identify your organism, but you can propose possible genera.
4. You might try different hydrocarbons to determine whether your organism can degrade other petroleum products. Does it grow on standard laboratory media? Can you think of other experiments using your isolated organism?

References

Nester et al. *Microbiology: A human perspective,* **6th ed. The McGraw-Hill Companies, 2009. Chapter 31, Section 31.4.**
Rosenberg, E. *The hydrocarbon oxidizing bacteria in the Prokaryotes,* 2nd ed., Vol. I, pgs. 446–456. Albert Balows et al. eds. New York: Springer-Verlag, 1992.

Figure 35.1 The hydrocarbon is placed on the filter paper in the inverted petri dish.

Agar

Filter paper

INDIVIDUAL PROJECT 36

Luminescent Bacteria: Bacteria That Produce Light

Getting Started

Some marine bacteria can emit light, a feature called *bioluminescence*. Most of these organisms live in relationship with tropical fishes. In one group of fish, a specialized organ near the eye supports the growth of these bacteria. By covering and uncovering the organ—and so the bacteria—with a special lid, the fish signal other fish in the darkness of deep water. Angler fish have an organ for *luminescent* bacteria that they dangle in front of them like a lure. Smaller fish are attracted to the light and are eaten. In the northern waters, the fish do not have special organs for luminescent bacteria. Here, these bacteria are found in fish intestines, are associated with squid, or are free-living in salt water.

The enzyme bacterial luciferase requires both oxygen and a protein called an inducer. One organism alone cannot produce enough of the inducer to permit light production—an entire population of bacteria is needed. This phenomenon is called *quorum sensing* and has recently been found in many other bacterial systems. Light production is very energy intensive—the equivalent of 6–60 molecules of ATP are needed for each photon of light produced. This system prevents one single bacterium from wasting energy to produce light that cannot be detected.

It is clear that luminescence may be an advantage to bacteria living symbiotically with fish. In exchange for light, they have a protected and rich environment. As previously mentioned, squid also may have special organs associated with these bacteria. Some squid store bacteria in a sac and expel the bacteria when attacked, similar to a cloud of ink. Others may use the bacteria to produce light so that they do not cast a shadow on bright moonlit nights, thereby protecting themselves from predators.

More puzzling is the presence of luminescent bacteria living in the gut of fish. Since this is an anaerobic environment, no light is produced, even though there is sufficient inducer. One proposal, which has some evidence to support it, is that the bacteria prefer the rich environment of the intestine, and when cast out into the water, they try to return by producing light and enticing the fish to swallow them.

Luminescent bacteria are also found free-living in salt water. If they grow on a piece of detritus or organic material, they might accumulate enough inducer to emit light. In this case, a population of bacteria is more successful than a single cell.

Many marine dinoflagellates luminesce when the salt water is warm. They frequently can be seen at night when water is agitated by waves lapping on the beach or by an anchor thrown into the water. It is not known what role luminescence plays in their ecology.

Procedure for Isolating Luminescent Bacteria

1. Prepare several plates of seawater complete agar (table 36.1). No one universal medium supports the growth of all luminescent bacteria, but this has been very successful.

Table 36.1	Seawater Complete Agar
Seawater	750 ml
Glycerol	3.0 ml
Peptone	5.0 grams
Yeast extract	0.5 grams
Agar	15.0 grams
Water	250 ml
pH 7.5	

Note: The seawater can be actual seawater or any kind of artificial seawater made up in the lab or purchased from an aquarium supply store. (From K. H. Nealson)

2. Obtain a whole saltwater fish or purchase a squid at a seafood store. Fresh squid is best, but frozen is also good if it is not thawed with hot water. Results with frozen fish have not been very successful.

3. Dip a swab in the intestinal content of the fish, and swab a third of the agar plate. Similarly, cut the squid open and try swabbing various organs. It is not clear where the luminescent bacteria are in highest concentration, but try swabbing the outside surface and the interior of the squid. Dispose of the swab in a wastebucket, and continue streaking for isolation with a loop.

4. Incubate plates at room temperature for 12–48 hours.

5. Observe for luminescent colonies. Take the plate into a dark room and permit your eyes to adjust for a few minutes. Circle any glowing colonies on the bottom of the plate with a marking pen. Try to look at the plate frequently in the first 24 hours because sometimes the bacteria are luminous for only a few hours. If your first attempt is unsuccessful, try again. Not all squid or fish have luminescent bacteria or at least ones that grow on this medium. If you leave the fish or squid unwrapped at room temperature,

you will sometimes see glowing colonies on their surface the next day.

6. Restreak a glowing colony onto another plate to obtain a pure culture. The organism is probably either *Photobacterium* or *Vibrio*.

7. If you would like to demonstrate the luminescence of your organisms to the class, you can write a message or draw an image on an agar plate with a swab dipped in your pure culture.

References

Bergey's Manual of Systematic Bacteriology, Vol. 1. Edited by Krieg, Noel, and John G. Holt, Baltimore/London: Williams & Wilkens, 1984.

Nealson, K. H. "Isolation, identification and manipulation of luminous bacteria," pp. 153–166 in *Methods in enzymology*, Vol. 57. Edited by M. Delucca. New York: Academic Press, 1978.

Nester et al. *Microbiology: A human perspective*, 6th ed. The McGraw-Hill Companies, 2009. Chapter 11, Section 11.7.

APPENDIX 1

Living Microorganisms (Bacteria, Fungi, Protozoa, and Helminths) Chosen for Study in This Manual

Acinetobacter, aerobic Gram-negative rods or coccobacilli in pairs. Commonly found in soil. Low virulence, can be an opportunist pathogen. Naturally competent and therefore easily transformed.

Amoeba proteus, a unicellular protozoan that moves by extending pseudopodia.

Aspergillus niger, a filamentous black fungus with a foot cell, columella, and conidia.

Bacillus cereus, a Gram-positive rod, forms endospores, found in soil.

Bacillus subtilis, a Gram-positive rod, forms endospores, found in soil.

Clostridium sporogenes, a Gram-positive rod, forms endospores, obligate anaerobe, found in soil.

Diphtheroids, Gram-positive irregular club-shaped rods.

Dugesia, a free-living flatworm.

Enterobacter aerogenes, a Gram-negative rod, coliform group, found in soil and water.

Enterococcus faecalis, a Gram-positive coccus, grows in chains, found in the intestinal tract of animals, occasionally an opportunist pathogen.

Escherichia coli, a Gram-negative rod, facultative, found in the intestinal tract of animals, coliform group, can cause diarrhea and serious kidney disease.

Escherichia coli, K-12 strain, commonly used in research. Host strain for λ phage.

Halobacterium salinarium, member of the Archaea, can live only in high salt solutions, where they can utilize chemical energy from organic material but can also obtain energy from light.

Klebsiella pneumoniae, Gram-negative rod, coliform group, opportunist pathogen.

Micrococcus luteus, Gram-positive obligate aerobe, cocci are arranged in packets of four or eight. Part of the normal biota of the skin. The yellow colonies are frequently seen as an air contaminant.

Moraxella, Gram-negative coccus found in normal biota.

Mycobacterium smegmatis, acid-fast rod with a Gram-positive type of cell wall.

Paramecium, a ciliated protozoan often found in pond water.

Penicillium species, a filamentous fungus with metulae, sterigmata, and conidia; also a source of antibiotics.

Propionibacterium acnes, Gram-positive irregular rod, obligate anaerobe, component of the normal skin biota living in sebaceous glands.

Proteus, Gram-negative rod, swarms on agar, hydrolyzes urea, can cause urinary tract infections.

Pseudomonas aeruginosa, Gram-negative rod, obligate aerobe, motile, opportunist pathogen, can degrade a wide variety of compounds.

Rhizopus nigricans, a filamentous fungus with stolons, coenocytic hyphae, and a sporangium containing asexual sporangiospores.

Saccharomyces cerevisiae, eukaryotic fungal yeast cell, replicates by budding. Important in bread, beer, and wine making and in the study of fungus genetics.

Spirillum volutans, Gram-negative curved organism with tufts of flagella at each pole, found in pond water and hay infusions.

Staphylococcus aureus, Gram-positive coccus, a component of the normal skin biota, but can cause wound infections, food poisoning, and toxic shock syndrome.

Staphylococcus epidermidis, Gram-positive coccus, a component of the normal skin biota.

Streptococcus mutans, normal biota of the mouth, forms gummy colonies when growing on sucrose.

Streptococcus pneumoniae, lancet-shaped Gram-positive cells arranged in pairs and short chains, pathogenic strains form capsules.

Streptococcus pyogenes, Gram-positive cocci in chains, the cause of strep throat, rheumatic fever, and glomerulonephritis.

APPENDIX

Dilution Practice Problems

See exercise 8 for an explanation of making and using dilutions.

1. If a broth contained 4.3×10^2 org/ml, about how many colonies would you expect to count if you plated
   a. 1.0 ml?
   b. 0.1 ml?
2. Show three ways for making each
   a. 1/100 or 10^{-2} dilution,
   b. 1/10 or 10^{-1} dilution, and
   c. 1/5 or 2×10^{-1} dilution.
3. Show two ways of obtaining a 10^{-3} dilution using 9.0 ml and 9.9 ml dilution blanks.
4. The diagram below shows a scheme for diluting yogurt before making plate counts. 0.1 ml was plated on duplicate plates from tubes B, C, and D. The numbers in the circles represent plate counts after incubation.

a. Which plates were in the correct range for accurate counting?
b. What is the average of the plates?
c. What is the total dilution of the tubes:

   A _____

   B _____

   C _____

   D _____

d. How many organisms/ml were in the original sample of yogurt?
5. Suppose an overnight culture of *E. coli* has 2×10^9 cells/ml. How would you dilute it so that you have countable plates? Diagram the scheme.

Answers

1a. 430 colonies

1b. 43 colonies

2a. 1.0 ml into 99 ml
0.1 ml into 9.9 ml
10 ml into 990 ml

2b. 1.0 ml into 9.0 ml
0.1 ml into 0.9 ml
10 ml into 90 ml

2c. 1.0 ml into 4 ml
0.1 ml into 0.4 ml
10 ml into 40 ml

3.

4a. Plates from tube C

4b. 55

4c. A: 10^{-2}
B: 10^{-4}
C: 10^{-5}
D: 10^{-6}

4d. The number of colonies $\times$ 1/dilution $\times$ 1/sample on plate $=$ number of organisms/ml

$55 \times 1/10^{-5} \times 1/0.1 = 55 \times 10^5 \times 10 = 55 \times 10^6$ or 5.5×10^7 org/ml

5.

APPENDIX 3

Use of the Metric System with Conversions to the English System of Measurement

The metric system enjoys widespread use throughout the world in both the sciences and nonsciences. Your studies in microbiology provide an excellent opportunity to learn how to use the metric system, particularly in the laboratory. The derivation of the word *metric* comes from the word *meter,* a measure of length.

The metric system is used in your microbiology course for media preparation, culture storage, and the measurement of cell number and size. The four basic measurements of the metric system are weight, length, volume, and temperature. The beauty of using the metric system is that all of these measurements are made in units based on multiples of 10. Not so with the English system, where measurements are made in units based on different multiples: 16 ounces in a pound, 12 inches in a foot, 2 pints in a quart, and 212°F, the temperature at which water boils, for example. The metric system weight measures in grams (g), length in meters (m), volume in milliliters (ml), and temperature in degrees Celsius (°C).

The prefixes of metric measurements indicate the multiple. The most common prefixes for metric measures used in microbiology are:

$$\text{kilo (k)} = 10^3 = 1{,}000$$

$$\text{centi (c)} = 10^{-2} = 0.01$$

$$\text{milli (m)} = 10^{-3} = 0.001$$

$$\text{micro (}\mu\text{)} = 10^{-6} = 0.000001$$

$$\text{nano (n)} = 10^{-9} = 0.000000001$$

Unit Only	Abbreviation	Equivalent
Weight		
kilogram	kg	1,000 g, 10^3 g
gram	g	1,000 mg (0.035 ounces; 454 g = 1 pound)
milligram	mg	10^{-3} g
microgram	μg	10^{-6} g
nanogram	ng	10^{-9} g
picogram	pg	10^{-12} g
Length		
kilometer	km	1,000 m (0.62 miles)
meter	m	100 cm (3.3 feet)
centimeter	cm	10^{-2} m (2.5 cm = 1 inch)
millimeter	mm	10^{-3} m
micrometer	μm	10^{-6} m (synonym = micron)
nanometer	nm	10^{-9} m
angstrom	Å	10^{-10} m
Volume		
liter	l	1,000 ml (1.1 quarts; 3.8 liters = 1 gallon)
milliliter	ml	10^{-3} liter
microliter	μl	10^{-6} liter
Temperature		
Celsius	°C	0°C = 32°F, 100°C = 212°F To convert from °C to °F: (°C × 9/5) + 32 To convert from °F to °C: (°F × 5/9) − 32 or 1.8°C = °F − 32

APPENDIX 4

Alternative Procedures

A. Spread Plate Technique

An alternative to making plate counts as described in exercise 8 is the spread plate technique. Instead of putting the sample in melted agar, it is placed on the surface of an agar plate and spread around with a bent glass rod (which resembles a hockey stick). Usually the sample size is 0.1 ml, so the following is a dilution scheme for suspension A and suspension B.

First Session

Suspension A

1. Label all water blanks with the dilution: 10^{-2}, 10^{-3}, 10^{-4}, 10^{-5}, 10^{-6}, 10^{-7}, 10^{-8}. Also label four petri dishes (on the bottom): 10^{-5}, 10^{-6}, 10^{-7}, 10^{-8}.
2. Make serial dilutions of the bacterial suspension A:
   a. Mix the bacterial suspension by rotating between the hands, and transfer 1.0 ml of the suspension to the 99-ml water blank labeled 10^{-2}. Discard the pipet.
   b. Mix well, and transfer 1.0 ml of the 10^{-2} dilution to the 9.0-ml water blank labeled 10^{-3}. Discard the pipet.
   c. Mix well, and transfer 1.0 ml of the 10^{-3} dilution to the 9.0-ml water blank labeled 10^{-4}. Discard the pipet.
   d. Mix well, and transfer 1.0 ml of the 10^{-4} dilution to the 9.0-ml water blank labeled 10^{-5}. Discard the pipet.
   e. Mix well, and transfer 1.0 ml of the 10^{-5} dilution to the 9.0-ml water blank labeled 10^{-6}. Discard the pipet.
   f. Mix well, and transfer 1.0 ml of the 10^{-6} dilution to the 9.0-ml water blank labeled 10^{-7}. Discard the pipet.
   g. Mix well, and transfer 1.0 ml of the 10^{-7} dilution to the 9.0-ml water blank labeled 10^{-8}. Discard the pipet.
   h. Mix well.

3. Starting with the most dilute suspension (10^{-8}), remove 0.1 ml and place on the surface of the agar plate. Immediately spread it around the entire surface of the plate with a sterile bent glass rod.
4. Using the same pipet and sterile glass rod (be careful not to contaminate), repeat the procedure for the 10^{-7}, 10^{-6}, and 10^{-5} dilutions.
5. Discard the pipet, and place the glass rod in a container as indicated by the instructor.
6. Invert the plates and incubate at 37°C.

Suspension B

1. Label all water blanks with the dilution: 10^{-2}, 10^{-3}, 10^{-4}, 10^{-5}, 10^{-6}. Also label four petri dishes (on the bottom): 10^{-3}, 10^{-4}, 10^{-5}, 10^{-6}.
2. Make serial dilutions of the bacterial suspension B:
   a. Mix the bacterial suspension by rotating between the hands, and transfer 1.0 ml of the suspension to the 99-ml water blank labeled 10^{-2}. Discard the pipet.
   b. Mix well, and transfer 1.0 ml of the 10^{-2} dilution to the 9.0-ml water blank labeled 10^{-3}. Discard the pipet.
   c. Mix well, and transfer 1.0 ml of the 10^{-3} dilution to the 9.0-ml water blank labeled 10^{-4}. Discard the pipet.
   d. Mix well, and transfer 1.0 ml of the 10^{-4} dilution to the 9.0-ml water blank labeled 10^{-5}. Discard the pipet.
   e. Mix well, and transfer 1.0 ml of the 10^{-5} dilution to the 9.0-ml water blank labeled 10^{-6}. Discard the pipet.
3. Starting with the most dilute suspension (10^{-6}), remove 0.1 ml and place on the surface of the agar plate. Immediately spread it around the entire surface of the plate with a sterile bent glass rod.

4. Using the same pipet and sterile glass rod (be careful not to contaminate), repeat the procedure for the 10^{-5}, 10^{-4}, and 10^{-3} dilutions.
5. Discard the pipet, and place the glass rod in a container as indicated by the instructor.
6. Invert the plates and incubate at 37°C.

Second Session for Both A and B Suspension (See pages 56–57)

Note: In the spread plate procedure, all the samples are 0.1 ml. This appears as $\frac{1}{0.1}$ for the sample size in the formula:

$$\text{number of organisms/ml in original sample} = \text{the number of colonies on plate} \times \frac{1}{0.1} \times \frac{1}{\text{dilution}}$$

B. An alternate method of calculating concentrations of organisms from dilution plate counts.

1. Convert the dilution to the dilution factor, which is the reciprocal of the dilution. For example, the dilution factor of a 10^{-1} dilution is 10. Convert all dilutions and samples to dilution factors, and then multiply by the average number of colonies on a countable plate.

2. For example: On page 57, the total dilution of the tubes in the box would be $10 \times 10 = 100$.

If a 1.0 ml sample from tube B contained 48 bacteria, the concentration in the original solution (not shown) would be $48 \times 10 \times 10$ organisms/ml or 48×100 organisms/ml.

3. Using the same example, if 0.1 ml were sampled, the reciprocal of 1/.1 is 10. The concentration would be $10 \times 10 \times 10 \times 48$ organisms/ml or 48×10^3/ml.

C. An alternative test for coagulase. The slide coagulase test is described in exercise 22. An alternative tube coagulase test involves the following steps. Mix a loopful of cells with 0.5 ml undiluted rabbit plasma in a small tube. Incubate at 37°C, and examine after 4–24 hours. The plasma can also be diluted 1:4 with saline. A solid clot is a positive test. This requires more reagent and takes longer, but some find it easier to read. See plate 26.

APPENDIX 5

Use of the Ocular Micrometer for Measurement of Relative and Absolute Cell Size

Determination of cell dimensions is often used in microbiology, where it has numerous applications. Examples include measuring changes in cell size during the growth cycle, determining the effect of various growth factors on cell size, and serving as a taxonomic assist in culture identification. Measurements are made by inserting a glass disc with inscribed graduations (figure 1a), called an ocular micrometer, into the ocular of the microscope (figure 2).

It is not necessary to calibrate the ocular micrometer to determine the *relative* size of cells, for this purpose, either a wet mount or stained preparation of cells can be examined with the ocular micrometer. By measuring the length in terms of number of ocular micrometer divisions, you might conclude that cell X is twice as long as cell Y. For such purposes, determination of cell length in absolute terms—such as number of micrometers—is not necessary.

For determining the *absolute* size of cells, it is necessary to first measure the length in micrometers (μm) between two lines of the ocular micrometer. A stage micrometer with a scale measured in micrometers is needed for μm calibration. The stage micrometer scale (figure 1b) is such that the distance between two lines is 0.01 mm (equivalent to 10 μm). By superimposing the stage micrometer scale over the ocular micrometer scale, you can determine the absolute values in microns between two lines on the ocular micrometer scale. The absolute value obtained is also dependent on the objective lens used. For example, with the low-power objective lens, seven divisions on the ocular micrometer equal one division on the stage micrometer. Thus, with the low-power objective lens, one division on the ocular micrometer equals .01 mm/7, which equals 1.40 micrometers (μm).

Procedure for Insertion, Calibration, and Use of the Ocular Micrometer

1. Place a clean stage micrometer in the mechanical slide holder of the microscope stage.

2. Using the low-power objective lens, center and focus the stage micrometer.

3. Unscrew the top lens of the ocular to be used, and then carefully place the ocular micrometer with the engraved side down on the diaphragm inside the eyepiece tube (figure 2). Replacing the top lens of the ocular. **Note:** With some microscopes, the ocular micrometer is inserted in a retaining ring located at the base of the ocular.

4. To calibrate the ocular micrometer, rotate the ocular until the lines are superimposed over the lines of the stage micrometer.

5. Next, move the stage micrometer until the lines of the ocular and stage micrometer coincide at one end.

6. Now find a line on the ocular micrometer that coincides precisely with a line on the stage micrometer.

7. Determine the number of ocular micrometer divisions and stage micrometer divisions where the two lines coincide. An example of this step is shown in figure 1c.

A relatively large practice microorganism for determining average cell size, with both the low- and high-power objective lenses, is a yeast cell wet mount. You must first calibrate the ocular micrometer scale for both objective lenses, using the stage micrometer. Next determine the average cell size of a group of cells, in ocular micrometer units, using both the low- and high-power objective lenses. Attempt to measure the same group of cells with both lenses. You may find it easier to first measure them with the high-power objective lens, followed by the low-power objective lens. You will have mastered the technique if you obtain the same average cell size answer with both objective lenses.

Note: Care should be taken when inserting, and removing the ocular micrometer from the ocular. Before inserting, make certain the ocular micrometer is free of dust particles by cleaning both sides with lens paper moistened with a drop of lens cleaning solution. Install and remove only in an area free of air currents. After removing the ocular micrometer, reexamine it for dust particles. If present consult your instructor.

Figure 2 Location of the ocular and stage micrometers.

Figure 1 Calibration of the ocular micrometer.

(a) Ocular micrometer
The diameter (width) of the graduations in microns must be determined for each objective.

(b) Stage micrometer
The graduations are .01mm (10 mm) wide.

(c) Superimposing of ocular micrometer scale (x) over the stage micrometer scale (y)
Note that seven divisions of the ocular micrometer equal one division of the stage micrometer scale (.01 mm).

One division of x = $\dfrac{.01}{7}$

= .0014 mm = 1.4 mm

(d) Based on the Figure 2(c) calculations, what is the average length in microns of the rod shaped bacteria?

APPENDIX 6

Use of the Hemocytometer for Determining Total Cell Number in a Liquid Suspension

A method faster than the plate count (see exercise 8) for determining the total number of cells present in a liquid suspension is one in which the hemocytometer is used in conjunction with the microscope. With this method, an aliquot of suspended cells is introduced between a cover glass suspended on mounts above the hemocytometer counting chamber (figure 1). The liquid depth between the cover glass and the counting chamber is 0.1 millimeter (mm).

The counting chamber is divided into a series of small squares in which the smallest squares are 1/400 of a square mm (see the central large square of figure 2). Thus a square mm would contain 400 small squares. The central large square is surrounded by double lines in order to make it easier to visualize when counting cells.

The hemocytometer is difficult to use with small cells because it is relatively thick. It can only be used with the low- and high-power objective lenses, thereby making it difficult to distinguish individual small cells. For white blood cells, yeasts, and larger bacterial cells it is sometimes quite useful. It is used routinely for white blood cell counts

and often for following the course of cell growth and multiplication in a liquid medium. For learning purposes yeast is an excellent test organism.

1. Dilute a test tube suspension of yeast until clouding is barely visible with the naked eye. You may need to further dilute the sample if you find the individual cells too dense to count with the hemocytometer.
2. Wash the hemocytometer and hemocytometer cover glass with soapy water, rinse with distilled water, and dry the cover glass and hemocytometer counting surface with lens paper or Kimwipes. Make certain that all oily residues are removed from these areas.
3. Place the cover glass over the counting chamber area.
4. Tighten the test tube cap of the yeast suspension and shake thoroughly.
5. Using a Pasteur pipet or plastic dropper, remove approximately 0.3 ml. While controlling the flow with your forefinger, place the tip in the V-shaped indention of the counting chamber adjacent to the edge of the cover glass (figure 1a).

Figure 1 (a) Top view of a hemocytometer, showing sample introduction point. (b) Side view of hemocytometer, showing the cover glass resting on the cover glass mounting supports and the (0.1 mm) distance between the top counting surface of the hemocytometer and the underside of the cover glass.

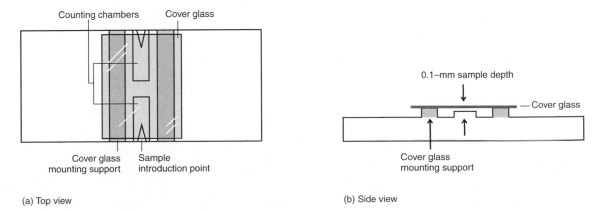

(a) Top view

(b) Side view

Figure 2 Ruling of a hemocytometer, showing the subdivisions of a central square millimeter. The central square is surrounded by double lines.

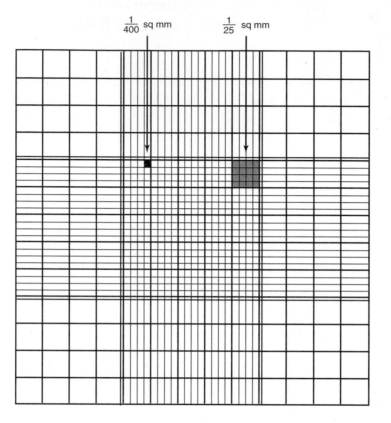

$\frac{1}{400}$ sq mm $\frac{1}{25}$ sq mm

6. Slowly let the counting chamber fill by capillarity, making sure that the suspension does not go between the cover glass and cover glass mounting supports of the counting chamber (figure 1*b*). Such an error will raise the height of the fluid under the cover glass which needs to be exactly 0.1 mm. If such an event occurs, return to step 2.

7. With care, place the hemocytometer on the stage of the microscope such that the hemocytometer counting chamber is centered underneath the low-power objective lens.

8. Focus and make a total count of the number of cells in a predetermined number of small squares of the double-lined central area of the hemocytometer, for example, 100 small squares. For ease in counting, a budding yeast cell should be treated as one cell. Assuming you find 500 cells in 100 small squares, what is the total number of cells per ml of sample?

9. Calculations:
   a. 100 small squares = 1/4 of a square mm; in which instance, 500 cells × 4 = 2,000 cells per square mm.
   b. As previously mentioned the liquid depth of the counting chamber is 0.1 mm. Thus in order to determine the number of cells per cubic mm of fluid, it becomes necessary next to multiply by a factor of 10 in order to obtain the number of cells per cubic mm. Thus 2,000 cells/sq mm × 10 = 20,000 cells per cubic mm.
   c. Finally in order to convert a cubic millimeter (mm) to a cubic milliliter (ml), it becomes necessary to multiply by a factor of 1,000 because 1,000 cu mm = 1 cu ml. Thereby 20,000 cells per cu mm × 1,000 = 20×10^6 yeast cells per ml of the original suspension.

APPENDIX 7

History and Working Principles of Light Microscopy

History

Anton van Leeuwenhoek (1632–1723), a Dutch linen draper, recorded the first observations of living microorganisms. He used a homemade microscope containing a single glass lens (figure 1) powerful enough to enable him to see what he described as little "animalcules" (now known as bacteria) in scrapings from his teeth and larger "animalcules" (now known as protozoa and algae) in droplets of pond water and hay infusions.

A single-lens microscope such as van Leeuwenhoek's had many disadvantages. These included distortion with increasing magnifying powers and a decreasing focal length (the distance between the specimen when in focus and the tip of the lens). Thus, when using a single lens with an increased magnifying power, van Leeuwenhoek had to practically push his eye into the lens in order to see anything.

Figure 1 Model of a van Leeuwenhoek microscope. The original was made in 1673 and could magnify the object being viewed almost 300 times. The object being viewed is brought into focus with the adjusting screws. This replica was made according to the directions given in *American Biology Teacher* 30:537, 1958. Note its small size. Photograph Courtesy of J.P. Dalmasso

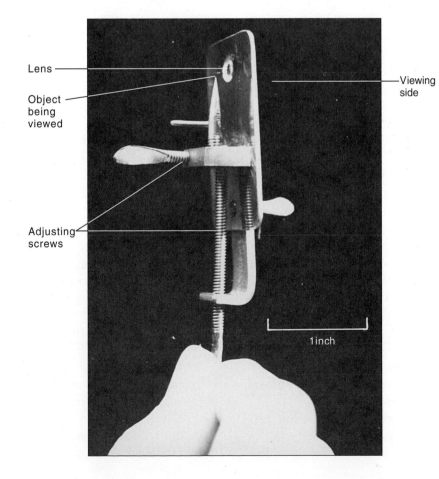

Figure 2 Modern bright-field compound microscope. Courtesy of Carl Zeiss, Inc.

Eyepiece (Ocular) a magnifying lens, usually about 10X

Objective nosepiece and attached objective lenses, each with a different magnification. The total magnification equals the product of the objective lens employed with the ocular lens

Specimen stage the platform on which the slide is placed

Iris diaphragm lever regulates the amount of light that enters the objective lens

Coarse-adjustment focusing knob

Fine-adjustment focusing knob

Condenser focuses the light

Base with illuminating light source

Modern microscopes, called **compound microscopes,** have two lenses: an eyepiece or ocular lens and an objective lens (figure 2). The eyepiece lens allows comfortable viewing of the specimen from a distance. It also has some magnification capability, usually 10 times (10×) or 20 times (20×). The purpose of the objective lens, which is located near the specimen, is to provide image magnification and clarity. Most teaching microscopes have three objective lenses with different powers of magnification (usually 10×, 45×, and 100×). Total magnification is obtained by multiplying the magnification of the ocular lens by the magnification of the objective lens. Thus, when using a 10× ocular lens with a 45× objective lens, the total magnification of the specimen image is 450 diameters.

Another giant in the early development of the microscope was a German physicist, Ernst Abbe, who (ca. 1883) developed various microscope improvements. One was the addition of a third lens, the **condenser lens,** which is located below the microscope stage (see figure 2). By moving this lens up or down, it is possible to concentrate (intensify) the light emanating from the light source on the bottom side of the specimen slide. (The specimen is located on the top surface of the slide.)

Abbe also developed the technique of using lens immersion oil in place of water as a medium for transmission of light rays from the specimen to the lens of the **oil immersion objective lens.** Oil with a density more akin to the microscope lens than that of water helps to decrease the loss of transmitted light, which, in turn, increases image clarity. Finally, Abbe developed improved microscope objective lenses that were able to reduce both **chromatic and spherical lens aberrations.** His objective lenses had a concave lens (glass bent inward like a dish) in addition to the basic convex lens (glass bent outward). Such a combination diverges the peripheral rays of light only slightly to form an almost flat image. The earlier simple convex lenses produced distorted image shapes due to spherical lens aberrations and distorted image colors due to chromatic lens aberrations.

Spherical Lens Aberrations   These occur because light rays passing through the edge of a convex lens are bent more than light rays passing through the center. The simplest correction is the placement of a diaphragm below the lens so that only the center of the lens is used (locate the **iris diaphragm** in figure 2). Such aberrations can also be corrected by grinding the lenses in special ways.

Chromatic Lens Aberrations   These occur because light is refracted (bent) as well as dispersed by a lens. The blue components of light are bent more than the red components. Consequently, the blue light travels a shorter distance through the lens before converging to form a blue image. The red components, travel a longer distance before converging to form a red image. When these two images are seen in front view, the central area, in which all the colors are superimposed, maintains a white appearance. The red image, which is larger than the blue image, projects beyond the central area, forming red edges outside of the central white image. Correction of a chromatic aberration is much more difficult than correction of a spherical aberration since dispersion differs among kinds of glass. Objective lenses free of spherical and chromatic aberrations, known as **apochromatic objectives,** are now available but are also considerably more expensive than **achromatic objectives.**

Some Working Principles of Light Microscopy

Microscope Objectives: The Heart of the Microscope

All the other parts of the microscope are involved in helping the objective lens attain a noteworthy image. Such an image is not necessarily the *largest,* but it is the *clearest.* A clear image helps achieve a better understanding of specimen structure. Size alone does not help achieve this end. The ability of the microscope to reveal specimen structure is termed **resolution,** whereas the ability of the microscope to increase specimen size is termed **magnification.**

Resolution, or resolving power, is also defined as the ability of an objective lens to distinguish two nearby points as distinct and separate. The maximum resolving power of the human eye when reading is 0.1 mm (100 μm). We now know that the maximum resolving power of the light microscope is approximately 0.2 μm, or 500× better than the human eye, and that it is dependent on the wavelength (λ) of light used for illumination as well as the numerical apertures (NA) of the objective and condenser lens systems. These are related in this equation:

$$\text{resolving power (r)} = \frac{\lambda}{NA_{obj} + NA_{cond}}$$

Examining this equation, we can see that the resolving power can be increased by decreasing the wavelength and by increasing the **numerical aperture.** Blue light affords a better resolving power

than red light because its wavelength is considerably shorter. However, because the range of the visible light spectrum is rather narrow, increasing the resolution by decreasing the wavelength is of limited use. Thus, the greatest boost to the resolving power is attained by increasing the numerical aperture of the condenser and objective lens systems.

By definition, the numerical aperture equals n sin theta. The **refractive index,** n, refers to the medium employed between the objective lens and the upper slide surface as well as the medium employed between the lower slide surface and the condenser lens. With the low- and high-power objective lenses, the medium is air, which has a refractive index of 1, whereas with the oil immersion objective lens, the medium is oil, which has a refractive index of 1.25 or 1.56. Sin theta is the maximum angle formed by the light rays coming from the condenser lens and passing through the specimen into the front objective lens.

Ideally, the numerical aperture of the condenser lens should be as large as the numerical aperture of the objective lens, or the latter is reduced, resulting in reduced resolution. Practically, however, the condenser lens numerical aperture is somewhat less because the condenser iris has to be closed partially in order to avoid glare. It is also important to remember that the numerical aperture of the oil immersion objective lens depends upon the use of a dispersing medium with a refractive index greater than that of air ($n = 1$). This is achieved by using oil, which must be in contact with both the condenser lens (below the slide) and the objective lens (above the slide). When immersion oil is used on only one side of the slide, the maximum numerical aperture of the oil immersion objective is 1.25—almost the same as the refractive index of air.

Note: Oil should not be placed on the surface of the condenser lens unless your microscope contains an oil immersion-type condenser lens and your instructor authorizes its use.

Microscopes for bacteriological use are usually equipped with three objectives: 16 mm low power (10×), 4 mm high dry power (40 to 45×), and 1.8 mm oil immersion (100×). The desired objective is rotated into place by means of a revolving nosepiece (see figure 2). The millimeter number (16, 4, 1.8) refers to the **focal length** of each objective.

By definition, the focal length is the distance from the principal point of focus of the objective lens to the principal point of focus of the specimen. Practically speaking, it can be said that the shorter the focal length of the objective, the shorter the **working distance** (that is, the distance between the lens and the specimen) and the larger the opening of the condenser iris diaphragm required for proper illumination (figure 3).

The power of magnification of the three objectives—10×, 45×, and 96×—is inscribed on their sides (note that these values may vary somewhat depending upon the particular manufacturer's specifications). The total magnification is obtained by multiplying the magnification of the objective lens by the magnification of the ocular eyepiece. For example, the total magnification obtained with a 4 mm objective (45×) and a 10× ocular eyepiece is 45 × 10 = 450 diameters. The highest magnification is obtained with the oil immersion objective lens. The bottom tip lens of this objective is very small and admits little light, which is why the iris diaphragm of the condenser lens must be wide open and the light conserved by means of immersion oil. The oil fills the space between the object and the objective lens so light is not lost (refer to figure 4.1 for visual explanation).

Note: If the microscope is *parfocal*, a slide that is in focus in one power will be in focus in other powers.

Figure 3 Relationship between the working distance of the objective lens and the diameter of the opening of the condenser iris diaphragm. The larger the working distance, the smaller the opening of the iris diaphragm.

Microscope Illumination

Proper illumination is an integral part of microscopy. We cannot expect a first-class microscope to produce the best results when using a second-class illuminator. However, a first-class illuminator improves a second-class microscope almost beyond the imagination. A student microscope with only a mirror (no condenser lens) for illumination can be operated effectively by employing light from a gooseneck lamp containing a frosted or opalescent bulb. Illuminators consisting of a sheet of ground glass in front of a clear bulb are available, but they offer no advantage over a gooseneck lamp. Microscope mirrors are flat on one side and concave on the other. In the absence of a condenser lens, the concave side of the mirror should be used. Conversely, when a condenser lens is present, the flat side of the mirror should be used since condenser lenses accept only parallel rays of light and focus them on the slide.

Two or more condenser lenses are necessary for obtaining the desired numerical aperture. The Abbe condenser lens, which has a numerical aperture of 1.25, is most frequently used. The amount of light entering the objective lens is regulated by opening and closing the iris diaphragm, located between the condenser lens and the light source (see figure 2). When the oil immersion objective lens is used, the iris diaphragm is opened farther than when the high dry or low-power objective lenses are used. Focusing the light is controlled by raising or lowering the condenser lens using the condenser knob.

The mirror, condenser, and objective and ocular lenses must be kept clean to obtain optimal viewing. The ocular lenses are highly susceptible to etching from acids present in body sweat and should be cleaned after each use.

Definitions

Achromatic objective. A microscope objective lens in which the light emerging from the lens forms images practically free from prismatic colors.

Apochromatic objective. A microscope objective lens in which the light emerging from the lens forms images practically free from both spherical and chromatic aberrations.

Chromatic lens aberration. A distortion in the lens caused by the different refrangibilities of the colors in the visible spectrum.

Compound microscope. A microscope with more than one lens.

Condenser. A structure located below the microscope stage that contains a lens and iris diaphragm. It can be raised or lowered, and is used for concentrating and focusing light from the illumination source on the specimen.

Focal length. The distance from the principal point of a lens to the principal point of focus of the specimen.

Iris diaphragm. An adjustable opening that can be used to regulate the aperture of a lens.

Magnification. The ability of a microscope to increase specimen size.

Numerical aperture. A quantity that indicates the resolving power of an objective. It is numerically equal to the product of the index of refraction of the medium in front of the objective lens (n) and the sine of the angle that the most oblique light ray entering the objective lens makes with the optical axis.

Oil immersion objective lens. High power lens (usually 100×) which requires the immersion of the lens into a drop of oil.

Parfocal. Having a set of objectives so mounted on the microscope that they can be interchanged without having to appreciably vary the focus.

Refractive index. The ratio of the velocity of light in the first of two media to its velocity in the second medium as it passes from one medium into another medium with a different index of refraction.

Resolution. The smallest separation which two structural forms, *e.g.*, two adjacent cilia, must have in order to be distinguished optically as separate cilia.

Spherical lens aberration. An aberration caused by the spherical form of a lens that gives different focal lengths for central and marginal light rays.

Working distance. The distance between the tip of the objective lens when in focus and the slide specimen.

APPENDIX

8

Additional Reading

Alibek, Ken. *Biohazard: the chilling true story of the largest covert biological weapons program in the world-told from the inside by the man who ran it.* Dell, 1999.

Atlas, Ronald. *Many faces, many microbes.* ASM Press, 2000.

Beck, Raymond. *A chronology of microbiology in historical context.* ASM Press, 2000.

Brock, Thomas. *Robert Koch: a life in medicine and bacteriology.* ASM Press, 1999.

Cook, Noble David. *Born to die: diseases and new world conquest, 1492–1650.* Cambridge University Press, 1998.

Crosby, Alred W. *America's forgotten pandemic: the influenza of 1918,* 2nd ed. Cambridge University Press, 2003.

Eigen, Manfred. *Steps toward life: a perspective on evolution.* Oxford University Press, 1992.

Ewald, Paul M. *Evolution of infectious disease.* Oxford University Press, 1996.

Garrett, Laurie. *The coming plague: newly emerging diseases in a world out of balance.* New York: Farrar, Straus and Giroux, 1994.

Geison, Gerald L. *The private science of Louis Pasteur.* Princeton University Press, 1995.

Harris, Henry. *The birth of the cell.* Yale University Press, 1999.

Hotez, Peter. *Forgotten people, forgotten diseases: The neglected tropical diseases and their impact on global health and development.* ASM Press, 2008.

Hudler, George W. *Magical mushrooms, mischievious molds.* Princeton University Press, 1998.

Jones, James H. *Bad blood: the Tuskegee syphilis experiment.* Simon & Schuster, 1992.

Karlen, Arno. *Man and microbes: disease and plagues in history and modern times.* Touchstone, 1995.

Kolata, Gina. *Flu: the story of the great influenza pandemic of 1918 and the search for the virus that caused it.* Farrar, Straus and Giroux, 2001.

McCormick, Joseph B. et al. *Level 4: virus hunters of the CDC.* Atlanta: Turner Publishing, 1996.

Nuland, Sherwin B. *The doctors' plague: germs, childbed fever and the strange story of Ignac Semmelweis.* W. W. Norton, 2003.

Oldstone, Michael. *Viruses, plagues, and history.* Oxford University Press, 2000.

Peters, C. J., and Olshaker, M. *Virus hunter: thirty years of battling hot viruses around the world.* Anchor Books, 1997.

Postgate, John. *The outer reaches of life.* Cambridge University Press, 1994.

Reverby, Susan M. *Tuskegee's truths: rethinking the Tuskegee syphilis study.* University of North Carolina Press, 2000.

Richards, Peter. *The medieval leper and his northern heirs.* Barnes & Noble, 1995.

Roueche, Berton. *The medical detectives,* revised edition. Plume, 1991.

Ryan, Frank. *The forgotten plague: how the battle against tuberculosis was won-and-lost.* Little Brown, 1992.

Sayre, Anne. *Rosalind Franklin and DNA.* W. W. Norton, 2000.

Schaechter, Elio. *In the company of mushrooms: a biologists tale.* Harvard University Press, 1997.

Sherman, Irwin. *The power of plagues.* ASM Press, 2006.

Sherman, Irwin. *Twelve diseases that changed our world.* ASM Press, 2007.

Tucker, Jonathan, ed. *Toxic terror: assessing terrorist use of chemical and biological weapons.* MIT Press, 2000.

Wainwright, Milton. *Miracle cure, the story of penicillin and the golden age of antibiotics.* Blackwell, 1990.

Watts, Sheldon. *Epidemics and history: disease, power and imperialism.* Yale University Press, 1997.

Wharton, David A. *Life at the limits: organisms in extreme environments.* Cambridge University Press, 2002.

INDEX

Erythromycin, 103t
Escherichia coli, 199, 209t, 265, 267, 279
Essential metabolic pathway, 101, 103
Essential metabolite, 102, 104
Ethanol, 102t
Ethidium bromide, in gel electrophoresis, 280
Ethylene oxide gas, 102t
Euglena oscillatoria, 263f
Eukaryotes, 291
Exotoxins
    definition of, 231
    Streptococcus, 229
Eyepiece, 16, 16f

F
F+, 125, 126
F−, 125, 126
Facultative anaerobes, 63
Fecal pollution, 255, 256f, 263–276, 264f
Fermentation, 144, 146, 149, 200, 201
Fermentation pathways, 200f
Fermentation technique, multiple-tube, 265–267, 266f, 268–271
Fermentation tubes, 199–200
F factor, 125, 126
Filamentous fungi, 143, 149
Fission, of yeasts, 144, 149
Flagella, 23
Flagella stain, 41
Fleming, Alexander, 223
Flora, skin, 187–190
Flukes, 163
Focusing knobs, 16f
Foot cell, 145f, 149
Forensic science, 279
Formaldehyde, 102t
Fuel oil, bacterial degradation of, 293–294
Fungi
    classification of, 144t
    ELISA test for, 237–244
    filamentous, 143, 149
    microscopic identification of, 143–153
    nonfilamentous, 143, 149
    Ouchterlony immunodiffusion assay for, 241, 245–253
    overview of, 139
    as pathogens, 146–148
    solute concentrations and, 93
    types of, 143

G
Gates, Bill and Melinda, 255
Gel electrophoresis, 279–281, 279f
Gene conjugation, 111, 125–128
Generation, spontaneous, 78
Generation time, 67–71, 94
Gene regulation, 111, 133–135
Genetic recombination, 119–121
Genetic reversion, 125
Gene transcription, 133, 134
Gene transduction, 111
Gene transfer, 111
Gene transformation, 111, 119–121
Germ tubes, 144, 149
Giardia lamblia, 264, 264f
Giardiasis, 139–140, 141
Glomerulonephritis, 229, 231
Glucose salts agar, 49
Glutaraldehyde, 102t
Glycerol, 102t
Glycolytic pathway, 149
Gram, Christian, 35
Gram-negative bacteria, 35
Gram-positive bacteria, 35
Gram staining, 27, 35–37
Granulocytes, 217
Group A *Streptococcus*, 229, 229t, 231
Group B *Streptococcus*, 229, 229t, 230, 231
Group D *Streptococcus*, 229, 229t, 230, 231
Growth
    aerobic, 63–64
    anaerobic, 63–64
    control of, 75–76
    environment and, 61
    of fungi, 139
    maximization of, 75
    microbial, 47
    minimization of, 75
    osmotic pressure and, 93–95
    phases of, 67, 67f
    temperature and, 67
    time and incubation technique, 67–71
    ultraviolet light and, 87–88
Growth curve, 94

H
Halobacterium, 93
Halophilic bacteria, 93–94
Health and Human Services, Department of, 255

Heat sterilization, 77–81
Heavy metal compounds, 102
Helminths, 161–170
Hemoglobin, 218
Hemolysins, 193
Heterotrophs, 139
Hexachlorophene, 102t
Hfr strains, 125, 126
High frequency of recombination, 125, 126
Histoplasma capsulatum, 147t
Histoplasmosis, 149
HIV disease (AIDS), 237, 255
Horseradish peroxidase, in ELISA testing, 237–241
Host
    definitive, 167
    intermediate, 167
Hydrocarbon-degrading bacteria, 293–294
Hydrogen peroxide, 102t
Hypertonic, 93, 94
Hyphae, 143, 145f, 146f, 149
Hypotonic, 93, 94

I
Identification, 207
Immersion lens, 23–24
Immersion oil, 23
Immune response, 215
Immunoglobulin G (IgG), in ELISA testing, 239–241
Immunoglobulin M (IgM), in ELISA testing, 239–241
Immunology, 215–216
Inclusion bodies, 29
Incubation
    definition of, 9
    generation time and, 67–71
Index, refractive, 24
Indicator organisms, 264–265, 268
Indirect ELISA test, 237, 238f
Individual projects, 291
Indole, 201
Inducible enzymes, 133
Infectious disease, 255
Ingenuity, ecological, 161
Inhibition, competitive, 102, 103
Inhibition zone, 231
Innate immunity, 223
Inoculate/Inoculation, definition of, 3
Intermediate host, 167
Intestinal flora, 199–202

Virology, 140
Virulent phage, 178, 179
Virus(es), 140–141, 177–178
Virus X, identification of, 280–284
Visible spectrum, 88
Voges-Proskauer test, 200–201

W

Water
    bacteriological examination of, 263–276
    disease transmission via, 255, 256f, 263
    fecal contamination of, 255, 256f, 263–265, 263f, 264f
    indicator organisms in, 264–265, 268
    membrane filter technique for, 267, 268, 269f, 271–272
    most probable number test for, 265, 266f, 270, 271t
    multiple-tube fermentation technique for, 265–267, 266f, 268–271
    purification of, 263–264
    total number of microorganisms in, 264
Wet mount
    definition of, 17
    motility determination in, 23–24
    preparation of, 18, 18f
White blood cell(s), 215, 218
White blood cell staining, 217–220
WHO (World Health Organization), 255
Wild type organism, 113
World Health Organization (WHO), 255
Wright's stain, 217

Y

Yeast dimorphism, 149
Yeast infections, 148, 149
Yeasts, 94, 143, 144, 147t, 194
Yellowstone National Park, 61
York, John, 255

Z

Ziehl-Neelsen acid-fast stain, 37
Zinc oxide paste, 102t
Zinc salts of fatty acids, 102t
Zygomycetes, 143, 144t